Classical Mechanics

Second Edition

Classical Mechanics
Second Edition

From Newton to Einstein: A Modern Introduction

Martin W. McCall
Imperial College London, UK

WILEY
A John Wiley and Sons, Ltd., Publication

This edition first published 2011

Registered office
John Wiley & Sons Ltd, The Atrium, Southern Gate, Chichester, West Sussex, PO19 8SQ, United Kingdom

For details of our global editorial offices, for customer services and for information about how to apply for permission to reuse the copyright material in this book please see our website at www.wiley.com.

Reprinted in January 2011

Library of Congress Cataloging-in-Publication Data

McCall, Martin W.
Classical mechanics : from Newton to Einstein : a modern introduction / Martin W. McCall. – 2nd ed.
p. cm.
Summary: "Classical Mechanics provides a clear introduction to the subject, combining a user-friendly style with an authoritative approach, whilst requiring minimal prerequisite mathematics - only elementary calculus and simple vectors are presumed. The text starts with a careful look at Newton's Laws, before applying them in one dimension to oscillations and collisions. More advanced applications - including gravitational orbits, rigid body dynamics and mechanics in rotating frames - are deferred until after the limitations of Newton's laws have been highlighted through an exposition of Einstein's Special Relativity. Big problems that are tackled using elementary techniques include the stability of the Universe, a body falling from a great height under gravity and Foucault's pendulum. Many new problems are included together with a supplementary web link to the solutions manual."– Provided by publisher.
Summary: "Classical Mechanics will be a clear introduction to the subject, combining a user-friendly style with an authoritative approach, whilst requiring minimal prerequisite mathematics"– Provided by publisher.
Includes bibliographical references and index.
ISBN 978-0-470-71574-1 (hardback)
1. Mechanics. I. Title.
QC125.2.M385 2011
531–dc22

2010022396

A catalogue record for this book is available from the British Library.

ISBN 9780470715741 (Hbk) 9780470715727 (Pbk)

Typeset by the author.

MIX
Paper from responsible sources
FSC www.fsc.org FSC® C013604

The author examining a rare second edition of Newton's *Principia* at the *Specola Vaticana*, Castel Gandolfo, May 1999.

He is not eternity, or infinity, but eternal and infinite. He is not duration and space, but He endures and is present. He endures forever, and is everywhere present; and by existing always and everywhere he constitutes duration and space. . .And thus much concerning God; to discourse of whom from the appearances of things, does certainly belong to natural philosophy.

Isaac Newton, 1687.

Contents

Preface to Second Edition

One of the major problems of producing this second edition of *Classical Mechanics* was deciding what new material to include. The most natural way to extend the work was to cover Lagrangian and Hamiltonian mechanics. Although these techniques are certainly very important, they are rather advanced, and I was keen to maintain the elementary flavour of the first edition. Moreover, as I discovered when I taught these techniques to third year undergraduates, the number of problems that become accessible is rather small; the achievements of these methods is principally conceptual in, for example, paving a path towards quantum mechanics. In fact the only problem I could find that really illustrated the progression from Lagrangian to Hamiltonian mechanics is the nutation of a gyroscope. Another option was to include a chapter on four-vectors. Again, whilst interesting, I didn't feel that it was quite in keeping with the spirit of the first edition, which was to teach mechanics and special relativity ab inito to undergraduate students with minimal mathematics. In the end I decided to embellish what I already had in the first edition. So here, I have included a new section on a body free-falling a large distance under gravity, which I haven't seen in textbooks before. New material on collisions is included to show that snooker balls always scatter at 90°. When I reconsidered the contortions to make the discussion of rigid bodies rotating about a fixed axis 'simple', I decided that the labour of introducing the inertia tensor was not so great, and consequently the rotation of arbitrary bodies is now discussed in Chapter 8. The Foucault pendulum is now discussed in detail, together with the 'tennis racket theorem' which pulls together material on rigid bodies, stability and rotating frames rather nicely. Some mathematical extension has been necessary to accommodate these topics, and the brief Appendix of the first edition has now been significantly extended. The chapter summaries have been extended where necessary to include the new material.

The complete manuscript has been re-typeset in LaTex, and a number of figures have been redrawn. New problems have been included, and a comprehensive online solutions manual has been prepared (www.wiley.com/go/mccall). I also took the opportunity of correcting a few typographical errors.

Preface to First Edition

The tale, as Tolkien wrote in the preface to Lord of the Rings, grew in the telling. Having copped the highest profile undergraduate course in the Physics programme at Imperial College, I set about writing the lectures on my laptop. With everything available in software, I felt it would be a relatively simple task to cut and paste the material into a book, and duly contracted with John Wiley & Sons Ltd to produce a camera ready manuscript in a few months. Thus I became pregnant with my first literary child. Little did I understand the pangs of labour.

The course entitled 'Mechanics and Relativity' is given to incoming undergraduates. I would meet them in the first week of their arrival and finish my forty first and final lecture in about the middle of the second term. The varied level of mathematical preparation of the class of 200 students set special problems for designing the course. Some would be familiar with solving differential equations, whilst others had done very little. I decided to take a 'lowest common denominator' approach in which more or less everything was derived in the lectures. I didn't want it to degenerate into dry mathematical machination, though, so I devised geometrical arguments through which some interesting results, such as the instability of the Universe, could be derived. This then was the brief, to develop an interesting, rigorous course covering Newtonian mechanics and relativity, with minimal mathematical prerequisites. It all sounds like a contradiction in terms, but I gave it my best shot.

I decided not to consult any books, so this one has very few references. Everything herein has been produced many times over since the times of Newton and Einstein, and all I could hope for was that my approach could be individual and fresh to the reader. The bibliography, however, lists the books I recommended at the beginning of the course as being suitable supplementary texts. I should acknowledge, however, that the chapters on relativity were undoubtedly influenced by the book from which I learned the subject: the first edition of Taylor and Wheeler's 'Spacetime Physics'. But I wanted to tell the story my own way, and so I made a conscious effort to think everything through for myself. I hope my understanding was good enough!

The problems provided at the end of each chapter are taken from those given to the students as classworks and problem sheets. They invariably start with some easy, confidence building exercises, before developing towards harder problems and examination questions. The brief summaries at the end of each chapter are intended to give the most concise exposition for revision; personally I've never found such things to be particularly helpful, but they are there for those who do.

There are many who have helped me with this book. I would particularly like to mention Gilbert Satterthwaite who cast his critical eye over the manuscript, and Michael

Niblock, a student who endured my first rendering of the course, who provided a valuable student's perspective on some chapters. Neal Powell and Meilin Sancho colluded to produce Figures 2.3, 3.17, 3.23, 7.12, 8.3, 8.4 and 8.5 – thank you. Thank you also to Keith Butt for giving veterinary supervision of the cat experiment of Figure 8.10, and of course for permitting Charlie to perform in the first place. It is also a pleasure to acknowledge Professor David Websdale from whom I inherited the course. He and his predecessors have undoubtedly influenced the book, not least by allowing me to use the problems and exercises that were passed on to me at the beginning. I have adapted many of these, and if any mistakes have crept in as a consequence, then I am responsible.

I never told my wife that I was writing this book I thought it would be fun to send her and her teaching colleagues a copy to review on publication so I can't give the customary 'without her tireless assistance, etc., etc.' acknowledgement. Nevertheless, Lulu, you have helped immeasurably in this project through your patience, kindness and love.

The West Indian anthropologist and cricket writer C.L.R. James wrote: 'Anyone who has participated in an electoral campaign will have noticed how a speaker, eyes red from sleeplessness, and sagging with fatigue, will rapidly recover all his power at an uproarious welcome from an expectant crowd.'[1] Well, not surprisingly, the class of some 200 students never roared expectantly whenever I entered the lecture theatre (!), but I can vouch for the ephemeral recovery of concentration amidst sleep deprivation – it was shortly after the birth of our son. Life is marginally less stressful now, and I've used the space–time to write this book. I hope you like it.

[1] James, C.L.R. (1966), *Beyond a Boundary*, Stanley Paul and Co.

1

Newton's Laws

1.1 What is Mechanics?

Even those uninterested in physics seem to have an intuitive notion about why and how things move. If a ball flies through the air, it does so because we have projected it—a *force* has been applied to a *body* which impels it to *move*. Glossing deftly over any difficulties there may be in defining what the italicised words actually mean, a description of the sequence force–body–motion is, conventionally, what is meant by **mechanics**. Mechanics is about forces and motion as applied to bodies.

Some people further resolve mechanics into **kinematics** and **dynamics**. Kinematics is the description of motion in terms of its trajectory through space as time progresses, or technically as $\mathbf{r}(t)$ where $\mathbf{r}$ is the position vector of the (dimensionless) body at time t. Typically this trajectory is calculated by solving an equation of motion. Dynamics, on the other hand, relates changes in a body's motion to their causes, i.e. forces. Dynamics is therefore the 'why' of motion, a typical dynamical problem being to find the resultant force acting on a body. Personally I have never found the delineation of kinematics and dynamics as branches of mechanics particularly useful, and so these terms will be avoided in this book.

1.2 Mechanics as a Scientific Theory

So how can we quantify the relation between the forces that act on bodies and their resultant motions? Given the obvious immediacy of this problem to our everyday lives, it is not surprising that progress in this area, firstly by Galileo and then by Newton, heralded the first truly scientific theory. A fair question to ask at this stage is: 'What are the characteristics of a scientific theory?' Without becoming too philosophical about the issue, it can be said that a scientific[1] theory is a concise summary of scientific ideas describing the results of experiments. Furthermore, a good theory will have predictive power beyond the domain of experimental experience to date. In physics, the theory is usually expressed as a mathematical relationship between experimentally determinable quantities. This means that the acid test which all physical theories must pass is:

[1]The adjective 'scientific' is perhaps unnecessary as, in my view, any theory in the sense described here can only be applied in the scientific context. Thus, 'management theory', 'education theory' and similar trendy oxymorons do not constitute bodies of ideas of any meaningful predictive power, and have merely empirical status.

Classical Mechanics – Second Edition From Newton to Einstein: A Modern Introduction Martin W. McCall

A good theory must agree with experiment.

Famous physicists have apparently disagreed over this fundamental idea. Feynman, in a popular lecture, commented that it does not matter how elegant the mathematical structure of the theory; if it disagrees with experiment, then it must be rejected. Dirac, however, famously reflected that it is better to have beauty in the mathematical statement of the theory, than for it to agree with experiment. They are both right. At the crude level, any gross disagreement with experiment must signal the rejection of a theory. However, Dirac's keen insight was that the most profound physical theories invariably have a beautiful mathematical structure, and apparent disparity with experiments may be due to extraneous factors. Even in well-established theories like classical mechanics, their mathematical structure is still being explored in interesting ways.

1.3 Newtonian vs. Einsteinian Mechanics

For 300 years the bastion of mechanics were the laws enunciated by Isaac Newton. And yet, even now, when better theories are known, 99.9% of the time practising physicists can still use Newton's laws with confidence. Again, this tells us something about the nature of scientific theories, for a minimum requirement of a new theory is that it reproduces all the results of its predecessors in appropriate limiting cases. Niels Bohr codified this precisely as the so-called Correspondence Principle. Thus when, in 1905, Einstein revolutionised physics by carefully refining our notions of space, time and frames of reference, the new laws of mechanics had to reproduce those of Newton when the speeds under consideration were much less than the speed of light.

Why bother to study Newtonian mechanics then, when we have better theories? Why not start and finish with the very best theory of mechanics to date, our predecessors having carefully checked its consistency with previous theoretical descriptions? The simple answer is that we learn incrementally from our experience, and so until we have some exposure to speeds approaching those of light we simply have no intuition on which to guide our deliberations, thus making the better theory harder to learn. Once in a lecture, and to the amusement of the class, I inadvertently omitted the '$\times 10^8$' when writing down the speed of light. When my slip was pointed out, I remarked how intuitive the notions of special relativity would become if indeed the speed of light were the 'everyday' speed of 3 ms^{-1}.

However, I believe there is a deeper answer relating to what might be called 'nature's hierarchy of beauty'. Yes, Einstein's relativistic description is more symmetric and, if we take symmetry to be a measure of aesthetic appeal (this is probably what Dirac meant), more beautiful than classical mechanics, but that symmetry is simply not apparent when analysing 'everyday' situations when $v \ll c$. In fact, at low speeds Newtonian mechanics develops a completely distinct geometrical structure (in which time plays the role of a parameter rather than a coordinate), in such a way as to make Newtonian theory a simplifying, or clarifying, description of nature at this level. In this sense nature appears to have a natural hierarchy of simplicity, and a theory that is naturally suited to the description of phenomena at different levels of detail.

Whatever the reason, most students find Einstein's relativity harder to understand at first reading than a review of the Newtonian mechanics they began to learn at school and so, since

Newtonian mechanics is still a remarkably successful theory for most situations, that is where we will begin.

1.4 Newton's Laws

Let us state Newton's laws, not in the archaic language of the *Principia* (1687), but in modern terminology:[2]

1. A body remains at rest or moves with constant velocity, $\mathbf{v}$, when no external force acts on it:

$$\frac{d}{dt}\mathbf{v} = \mathbf{0} \ \Leftrightarrow \ \mathbf{F} = \mathbf{0}. \tag{1.1}$$

2. The rate of change of momentum of a body is proportional to the force on the body:

$$\mathbf{F} \propto \frac{d}{dt} m\mathbf{v} \,. \tag{1.2}$$

3. When two bodies interact, they exert on each other equal, but opposite, forces:

$$\mathbf{F}_{12} = -\mathbf{F}_{21} \,. \tag{1.3}$$

Now let us carefully unpack these statements seeking, as we go, their real content. Ask yourself the following: bearing in mind the brief discussion above on the nature of physical theories, are the statements laws, postulates, definitions or empirical observations?

The first law is Galileo's law of inertia and breaks away from the ideas of mechanics dating back to Aristotle that a body's natural state is at rest. The intellectual leap in formulating the first law should not be underestimated. We are mostly familiar with the notion that bodies slow down as they move, apparently striving to achieve, according to the Aristotelean view, a 'natural state of rest'. Of course we are now comfortable with the idea that bodies slow down because of the application of frictional forces, though this could hardly have been self-evident at the time of Galileo and Newton. If constant velocity is the 'natural' state of motion when a body is 'free' then interactions from external influences change this 'natural state'. The second law tells us how.

The second law (which in fact was first written down by Robert Hooke) is the kernel of Newtonian mechanics, quantifying the relationship between forces and motion, and asserting that the rate of change of a body's momentum is proportional to the force impelling a body to move. The definition that momentum is mass times the velocity is implicit. The constant of proportionality, k, is, by choice of units, defined to be equal to one, the resulting unit of force being equal to one newton (appropriately enough), usually written as 1 N. Thus a force of 1 N causes a mass of 1 kg to accelerate at $1\,\mathrm{m\,s^{-2}}$. The weight of an apple is (appropriately enough) about 1 N. We are taught at school to differentiate carefully between mass and weight, the latter being the force exerted on a mass due to gravity at the surface of

[2]Vector quantities, with which the reader is assumed to have a little familiarity, appear in bold face. However, when considering motion in one dimension vector quantities will appear in italic face, the direction being dictated by the sign. Note the distinction between the null vector $\mathbf{0}$, whose components all vanish, and the scalar 0.

a planet (usually the Earth). Most systems to which the second law is applied have constant mass (though rocket motion is an important exception – see Section 4.8), in which case

$$\mathbf{F} = m\frac{d\mathbf{v}}{dt} = m\mathbf{a} , \tag{1.4}$$

where $\mathbf{a}$ is the body's acceleration. This is the form in which the second law is most frequently stated, though it is very important to remember that it can only be written in this way for constant mass systems. Only then can we say that the first law is a special case of the second, since if $\mathbf{F} = \mathbf{0}$, then integrating Equation (1.4), gives,

$$\mathbf{v} = \mathbf{constant} , \tag{1.5}$$

which is the first law.

The third law codifies the relationship between forces acting between two bodies. If $\mathbf{F}_{12}$ is the force acting on body one due to body two, then Newton's third law asserts that body two exerts an equal and opposite force on body one, i.e. $\mathbf{F}_{12} = -\mathbf{F}_{21}$. Careful: never has there been an elementary law of physics so misunderstood. The third law does *not* refer to two forces acting on the *same* body. So when a book rests on a table, the force of gravity *on the book* is balanced by the electromagnetic forces which keep both bodies rigid, thus providing a reaction *on the book* (see Figure 1.1). Apply more downward force to the book and the

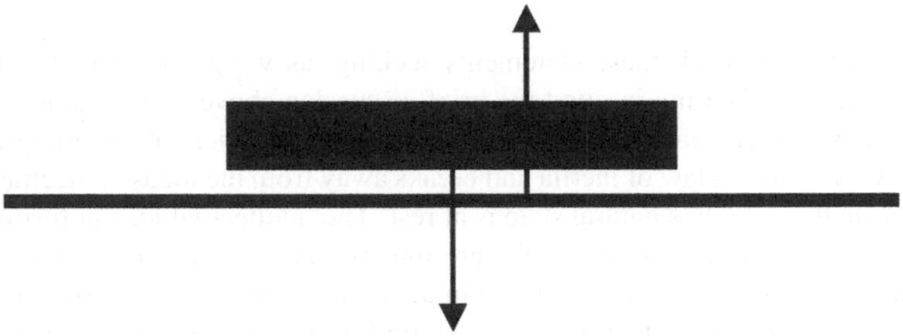

Figure 1.1 A book resting on a table: the forces acting on the book are *not* a good example of Newton's third law.

electromagnetic forces maintaining the rigidity of the system will continue to keep the book at rest until the table collapses. The third law applies to the reciprocal forces between two bodies interacting with each other. When two masses are connected by a spring the forces acting on each mass due to the other mass are equal and opposite (see Figure 1.2). This is a good example of the third law in action, which we will use shortly to measure mass. Avoid using the phrase 'Action and reaction are equal and opposite', which is at best unclear. A consequence of Newton's laws is that if $\mathbf{F}_{12} = -\mathbf{F}_{21}$ then, by the second law,

$$\frac{d}{dt}(m_1\mathbf{v}_1 + m_2\mathbf{v}_2) = \mathbf{0} , \tag{1.6}$$

or

$$m_1\mathbf{v}_1 + m_2\mathbf{v}_2 = \mathbf{constant} . \tag{1.7}$$

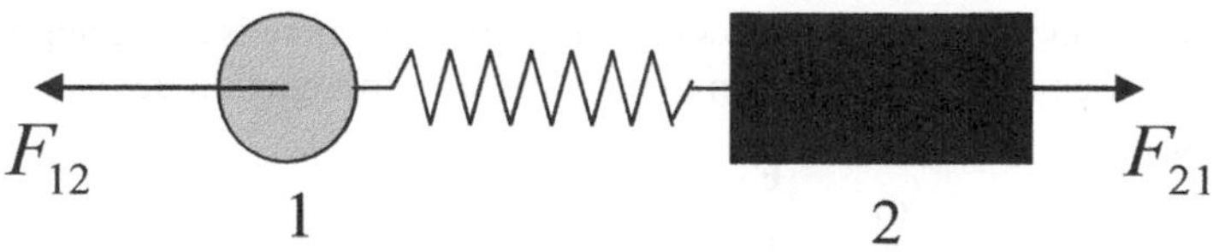

Figure 1.2 The mutual interaction of two masses connected by a spring as an example of Newton's third law.

Thus, for a two-body system in the absence of any external forces, the momentum is conserved. Clearly this idea can be extended to an arbitrary number of bodies so that we can assert that for an isolated system (i.e. one where no external forces are acting) of N bodies, the total momentum is conserved:

$$m_1\mathbf{v}_1 + m_2\mathbf{v}_2 + \cdots + m_N\mathbf{v}_N = \textbf{constant} . \tag{1.8}$$

Beware! Whenever a lecturer or author uses the phrase 'clearly it follows that so-and-so is true', almost invariably it conceals some extra subtleties. The extension of momentum conservation to a system of N bodies is explored more rigorously in Problem 1.1 at the end of this chapter.

1.5 A Deeper Look at Newton's Laws

Let us start by examining carefully the terms used in the statement of the laws. What is a *force*? It is actually rather hard to say what a force *is*, rather than what it *does*. What it does is produce acceleration (for constant mass systems), but how can we test the proportionality between force and acceleration, without some independent quantification of force? Maybe we could use Hooke's law in relation to spring extensions to measure forces, though we might accidentally exceed the elastic limit and obtain spurious results (which actually violate the second law). So it seems that the second 'law' is almost a kind of definition of force. To some extent this is true, since mechanics provides no prescription as to how forces arise. The forces are presumed to be 'given' from other branches of physics – for gravitation and electrostatics by inverse square laws, for example. Once a force law is given, the second law allows us to make further progress by relating how that force impels a body to move. Moreover, as shown above, in the absence of external forces, the momentum (*defined* as mv) is conserved for an isolated system. Although momentum has been defined as mv, the *prediction* of its conservation in isolated systems is testable by experiment. It is then the business of experimental physics to confirm that momentum for an isolated system is indeed conserved. If it is, then we have found out something fundamental about nature and the statements are beyond mere definitions. Momentum *is* found to be conserved experimentally.

What about mass? Mass measures the amount of matter present in a body. As implied by the second law, it measures how hard we have to push a body to achieve a given acceleration. But again we encounter the potentially circular reasoning whereby the second law is, initially at any rate, being used as a kind of definition of mass. To progress further, and to have some measure of mass that is somehow independent of the second law, we must subject two masses (a test mass and the mass we wish to measure, say) to the *same* force and compare their accelerations. How can we know that the same magnitude of force is being applied? Answer: use the third law! (See Figure 1.2.) From the second law, the magnitude of the acceleration of the two masses is

$$a_1 = \frac{|\mathbf{F}_{12}|}{m_1} \text{ and } a_2 = \frac{|\mathbf{F}_{21}|}{m_2} ,$$

so that the mass ratio is given by

$$\frac{m_2}{m_1} = \frac{|\mathbf{F}_{21}|/a_2}{|\mathbf{F}_{12}|/a_1} = \frac{a_1}{a_2}, \tag{1.9}$$

since $|\mathbf{F}_{12}| = |\mathbf{F}_{21}|$, by the third law. Hence, if the accelerations of each of the masses are measured as the spring pushes them apart, then the mass m_2 is given by

$$m_2 = m_1 \left(\frac{a_1}{a_2}\right) . \tag{1.10}$$

Equation (1.10) then measures the mass of an unknown test mass, m_2, in units of a standard mass, m_1.

In this context it is worth drawing the distinction between **inertial mass** and **gravitational mass**. What we have discussed above is *inertial* mass, being the mass used in Newton's second law. However, Newton's theory of gravitation states that the attraction between two bodies separated by a distance r is given by

$$\mathbf{F} = -\frac{G m_G M_G}{r^2} \hat{\mathbf{r}} , \tag{1.11}$$

where G is a universal constant and $\hat{\mathbf{r}}$ is a unit vector directed along the separation between the two masses, the minus sign indicating attraction. The subscript G on the two masses, m and M, indicates that there is no a priori connection between the mass appearing in Equation (1.11) and the inertial mass of the second law. Mass in the gravitation law is referred to as *gravitational* mass. It is the 'charge' associated with the gravitation law (the corresponding law for electrostatic attraction in vacuum is $\mathbf{F} = \left(Q_1 Q_2 / 4\pi\epsilon_0 r^2\right) \hat{\mathbf{r}}$ where Q_1 and Q_2 are the charges and ε_0 is a constant). Now it happens that experimentally it is found that the gravitational mass of a body is proportional to its inertial mass. Moreover, by appropriate choice of units, the gravitational mass is *equal* to the inertial mass. Galileo's experiment to show the proportionality is the most famous physics experiment of the post-Renaissance era. Whether Galileo actually dropped balls from the top of the Leaning Tower of Pisa (Figure 1.3) is debatable. What matters is whether he did the experiment at all, which undoubtedly he did. Applying Newton's second law to the two masses:

$$-\frac{G M_G^{(E)} m_G^{(1)}}{R_E^2} = m_I^{(1)} a_1 , \quad -\frac{G M_G^{(E)} m_G^{(2)}}{R_E^2} = m_I^{(2)} a_2 , \tag{1.12}$$

where $M_G^{(E)}$ is the (gravitational) mass of the Earth and R_E is the Earth's radius. Dividing Equations (1.12)

$$\frac{a_1}{a_2} = \left(\frac{m_G^{(1)} / m_I^{(1)}}{m_G^{(2)} / m_I^{(2)}}\right) = \left(\frac{m_G^{(1)} m_I^{(2)}}{m_G^{(2)} m_I^{(1)}}\right) . \tag{1.13}$$

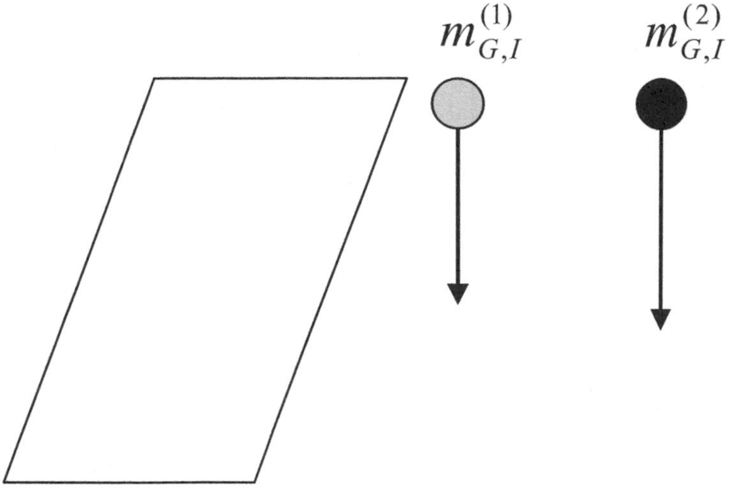

Figure 1.3 The Leaning Tower of Pisa experiment. Different masses are dropped from the same height. If they fall with the same acceleration then inertial mass is proportional to gravitational mass. Air resistance is neglected.

Thus *if* m_G/m_I is the same for all masses, then all bodies will fall with the same acceleration, and $a_1 = a_2 = -g$, the acceleration due to gravity. If not, then the bodies will separate as they fall. Modern experiments have tested the equality between gravitational and inertial mass to better than 1 part in 10^{12} and thus experimentally, $m_G/m_I =$ *constant* which, as noted above, can be set equal to one by choice of units, that is

$$m_G = m_I \ . \tag{1.14}$$

Of course it would be a remarkable coincidence if nature had simply decided that gravitational mass is to be proportional to inertial mass without any deeper significance. It suggests that gravitation can be described as a form of inertia, or, in the context of inertial frames, that gravitation is an inertial force (see below). A careful exposition of such issues is the basis of Einstein's general theory of relativity, which is beyond the scope of this book.

1.6 Inertial Frames

So far we have said nothing about the most fundamental assumptions in Newton's laws, concerning the nature of space and time. Space is presumed to follow the rules of geometry we learned at school, that is Euclidean geometry.[3] Although this may seem 'obvious', it is an underlying assumption for which we must supply an experimental test to verify its applicability. The appropriate test is to use Pythagoras' theorem. If Pythagoras' theorem holds for all right-angled triangles in three-dimensional space, then space is Euclidean, or 'flat'.

[3] An example of a *non*-Euclidean geometry is *spherical geometry* where 'lines' are great circles on the surface of a sphere. The angles of a spherical triangle sum to more than $180°$.

Newton regarded both space and time as 'absolute' entities, implying that they are the same for everyone. This immediately brings forth the notion of a *frame of reference*. For our purposes a reference frame may be considered as an imaginary latticework of rulers extending orthogonally in the three spatial directions. Via this frame the location of bodies may be determined. To complete the mechanical description, to each lattice is appended a clock that pinpoints when a particle is at a given location. Thus the trajectory through space as time evolves is determined in the frame of reference. Newton's notion of 'absolute time' is the tacit assumption that the time measured in one frame is the same as the time measured in any other. Time measured in two frames that are differently oriented, for example, is assumed to be the same[4]. This seemingly intuitive aspect of time fails dramatically when frames in relative motion are considered, but that will be the business of special relativity to be considered in Chapters 5 and 6. It is interesting to note, and it is a tribute to Newton's genius, that although he based his mechanics on absolute space and time he was actually uncomfortable with the rigidity that this imposed. However, and again in tribute to his genius, he recognised this as the only way to make practical progress at the time.

Even within the above definition of reference frames, there is still considerable latitude in choosing frames. Frames can be in motion relative to each other, and the question arises as to whether Newton's laws apply in all frames, or just some special class of frames. Newton's first law supplies the clue. If I hold a cup of coffee at my desk, or in the cabin of an aeroplane, I hardly notice any difference. Thus frames in *uniform* relative motion are apparently equivalent from the mechanical viewpoint. However, try the same experiment on a helter-skelter. Evidently, a frame that is accelerating with respect to another frame is not one in which bodies move uniformly and must be rejected. If we designate frames that pass the Newton's first law test as inertial frames, we are left with only one possible definition:

- An **inertial frame** is one in which Newton's first law is valid.

Yet again we must examine closely whether our reasoning is circular. First let us remove any extraneous external interactions by setting up our ideal inertial frame in deep space, far removed from other matter. We can decide whether or not our frame is rotating by looking at the motion of a 'free' body within the frame. If our record of the trajectory of the body shows it to be of a complex spiralling nature, then the frame is rotating. If, however, the motion is uniform, then our frame is inertial.[5] Such an ideal inertial frame may also be referred to as a **free-float frame** – see Figure 1.4.

All of this is seen to be a conceptual idealisation for which in practice successive approximations must be made. Let us consider a few examples:

1. *Deep space* – The cosmos exhibits a 'clumpiness' on a scale of about 150 million light years ($\sim 10^{25}$ m). By removing ourselves to a remote region, we may still have to think about the odd hydrogen atom, but this is the best possible approximation to the 'free-float' or 'ideal' inertial frame considered above. Such a frame is of interest only to cosmologists and astronomers.

[4]Of course the observers in different frames may set their time origins differently, so what is really meant here is the *time difference* between the same pair of events as recorded in different frames.

[5]There are additional subtleties here concerning the nature of time. If we have a 'bad' clock that measures time as a nonlinear function of the time shown by a 'good' clock, motion will appear nonuniform even in an inertial frame. But how do we know a priori whether our clock is good or bad?

2. *Frame of 'fixed stars' in our galaxy* – Over the orbital periods of the planets in our solar system, the stars of our galaxy show little motion, and furthermore we can suppose that the stars beyond the sun hardly influence the motions of the planets. Thus planetary motion may in practice be analysed in a frame tied to the 'fixed stars' of our galaxy.

Figure 1.4 A free-float frame: a region of space removed as far as possible from other matter.

3. *Earth* – The use of the Earth as an inertial frame presents certain difficulties. First, whilst the initial motion of a bullet fired from a gun might pass the inertial frame test, most motions we want to consider are over a sufficient duration to be significantly affected by the Earth's gravity. One way to proceed is to regard Earth gravity as an external force applied to the motion of all bodies within an Earth-bound inertial frame:

$$\mathbf{F}_{\text{grav}} = m\mathbf{a} = -mg\mathbf{k} \tag{1.15}$$

where the standard unit Cartesian basis vectors $(\mathbf{i}, \mathbf{j}, \mathbf{k})$ have been introduced, with the unit vector $\mathbf{k}$ pointing 'up'.

However, since we cannot switch gravity off (at least not directly, see below), we have no way of applying the Newton III test. At best we can verify Newton's third law in a plane orthogonal to the direction of gravity by counteracting its effects with a vertically applied force, for example a puck on a frictionless table, with the table supplying the upwards reaction to gravity.

Moreover, the fact that the Earth rotates means that it is an accelerating frame. The rotation rate, ω, is 2π radians every 24 hrs, or $\omega = 7.3 \times 10^{-5}\,\text{rad}\,\text{s}^{-1}$. As a rough guide, we can ignore any effects due to the Earth's rotation over times that are short compared with the rotation period, though we must bear in mind that rotational effects will always be detectable with sufficiently accurate measurements. Effects due to the

Earth's rotation are the centrifugal effect and the Coriolis effect, to be discussed in Chapter 9.

All of the above approximations to inertial frames are 'valid', in that it depends entirely on the accuracy required as to whether a given approximation to an inertial frame is adequate.

Ultimately nowhere is remote enough to be classed as an inertial frame—suggesting that it should be possible to formulate mechanics without this conceptual idealisation. Indeed this is possible, but it would lead us too far afield for this book. The difficulties in defining an ideal inertial frame strike right at the heart of Newton's system for describing mechanics.

Since it appears that for terrestrial experiments we are stuck with analysing mechanics in noninertial frames, it is useful to consider briefly whether we can salvage anything in terms of attempting to apply Newton's laws in them.

1.7 Newton's Laws in Noninertial Frames

Consider the situation shown in Figure 1.5 which depicts a plumb-bob suspended from the ceiling of a railway carriage which is accelerating to the right with acceleration $\mathbf{a}$. The string tension $\mathbf{T}$, and the bob weight mg, are the identified forces acting in the directions indicated. Now what is the effect of attempting to apply Newton's laws in the frame of the carriage? In this frame the bob is in equilibrium, so that the vector sum of the forces must cancel. What is there to oppose the horizontal component of the tension $T \sin\theta$? An observer in the inertial frame of the platform sees the train accelerate by a to the right and, applying Newton's second law, writes $T \sin\theta = ma$. In order for the train-based observer to obtain equilibrium he must invent a force $\mathbf{F}_I = -m\mathbf{a}$. The introduction of this force is necessary purely due to the attempt to use Newton's laws in an accelerating frame of reference.

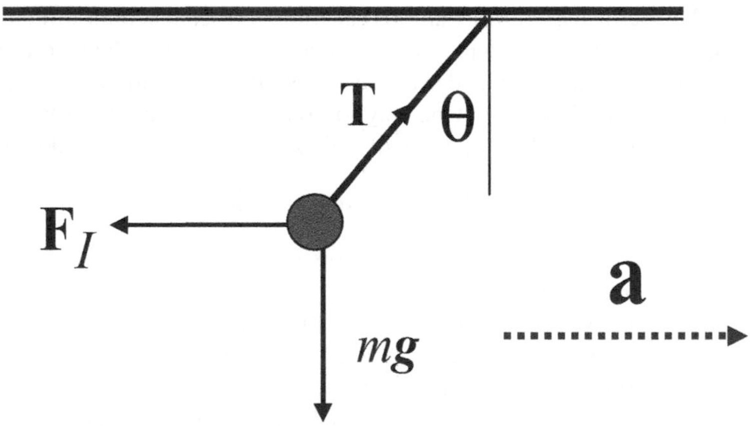

Figure 1.5 Plumb-bob in an accelerating railway carriage.

In fact this is a general result. Newton's laws may be applied in a frame accelerating with constant acceleration $\mathbf{a}$, provided that in addition to the identified forces acting, each body is subjected to a force

$$\mathbf{F}_I = -m\mathbf{a} \tag{1.16}$$

where m is the mass of the body. These forces are sometimes called **inertial forces** or, less helpfully, **fictitious forces**.

1.8 Switching Off Gravity

Now comes a beautiful trick. Can we combine the ideas of the previous two sections to manufacture a good inertial frame near the Earth? Let us analyse how mechanics looks to an individual in a lift cabin when the lift cable breaks and the cabin is in freefall – see Figure 1.6.

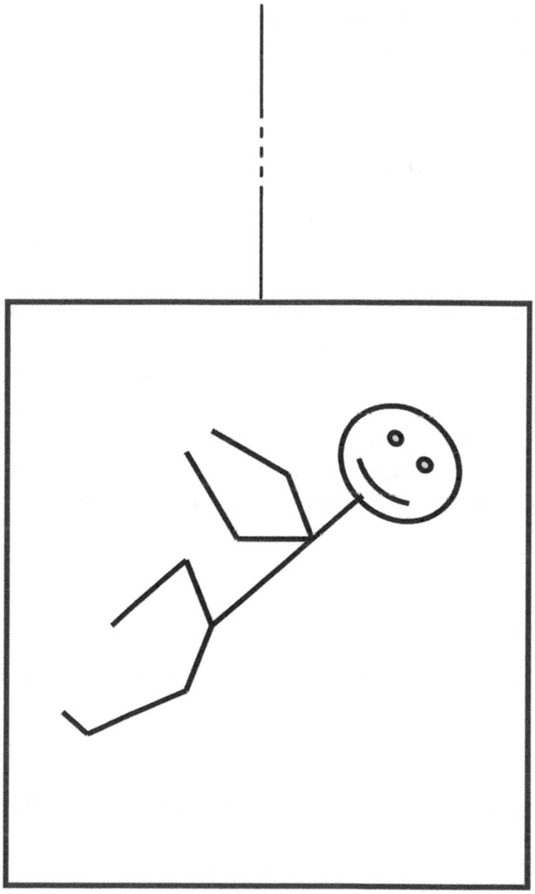

Figure 1.6 Free-fall lift cabin; despite his impending fate, the passenger is pleased that, for a while, he is in a good inertial frame.

The only force acting is gravity, which is given by $F_{\text{grav}} = mg$, taking downwards as positive. But the elevator frame is accelerating down with precisely the same acceleration, so that applying the prescription above, when analysing in this frame, we must introduce a force given by $F_I = -mg$. The net force on the individual, and any other free body in the cabin, is thus given by

$$F_{\text{grav}} + F_I = mg - mg = 0. \tag{1.17}$$

Any additional forces that may be acting between the bodies in the cabin can now safely be analysed using Newton's laws. We have thus succeeded in constructing a 'quasi'-inertial frame near a large gravitating mass (the Earth), in which Newton's laws can be applied. Gravity has been 'switched off'!

Having said that, we must recognise that the lift cabin can only occupy a small volume, and can only fall for a short time. If not, then so-called tidal effects become significant, and gravity can be detected again – see Problem 1.2 at the end of this chapter.

A profound insight from Einstein was to use the above ideas to postulate that since an inertial force can oppose gravity, then gravity itself may be described as an inertial force. Succinctly, uniform gravity is indistinguishable from acceleration. This line of reasoning also leads to a very natural explanation for the equality of inertial and gravitational mass, a complex of ideas known as the equivalence principle.

1.9 Finale – Laws, Postulates or Definitions?

Let us return to the question posed just after the statement of Newton's laws: are they laws, postulates, definitions or empirical observations? As we have seen their real nature is actually very hard to penetrate, as they contain elements of all four, though they are certainly more than any aspect individually. Perhaps the best we can do is to summarise and say that they constitute a body of ideas passing all the tests for a good physical theory: summary of observational data in mathematical statements, and prediction of results of experiments not yet performed. The crowning achievement of Newton's theory was its application beyond our everyday, terrestrial experience to describe lunar and planetary motion. Newton's theory was thus the first attempt at a universal, or *fundamental*, description of nature, being applied to the motion of *all* bodies subjected to forces of *every* possible origin.

1.10 Summary

Section 1.4. Newton's Laws:

1. A body remains at rest or moves with constant velocity, v, when no external force acts on it:
$$\frac{d}{dt}\mathbf{v} = \mathbf{0} \Leftrightarrow \mathbf{F} = \mathbf{0}\,. \tag{1.1}$$
2. The rate of change of momentum of a body is proportional to the force on the body:
$$\mathbf{F} \propto \frac{d}{dt}m\mathbf{v}\,. \tag{1.2}$$
3. When two bodies interact, they exert on each other equal, but opposite, forces:
$$\mathbf{F}_{12} = -\mathbf{F}_{21}\,. \tag{1.3}$$

The consequence of Equations (1.2) and (1.3) is that momentum is conserved for an isolated system:

$$\frac{d}{dt}(m_1\mathbf{v}_1 + m_2\mathbf{v}_2) = \mathbf{0} \qquad \Rightarrow \qquad m_1\mathbf{v}_1 + m_2\mathbf{v}_2 = \mathbf{constant}\ .$$

Equations (1.6) and (1.7).

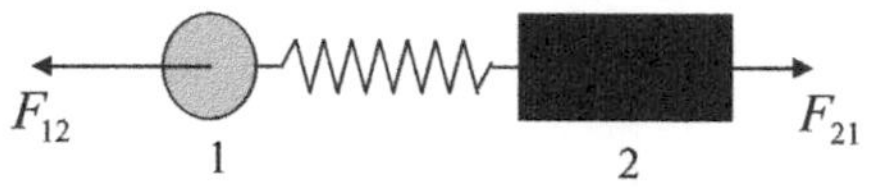

Figure 1.2

Section 1.5. Use Newton's third law to measure (inertial) mass:

$$\frac{m_2}{m_1} = \frac{|\mathbf{F}_{21}|/a_2}{|\mathbf{F}_{12}|/a_1} = \frac{a_1}{a_2}\ . \tag{1.9}$$

Distinction with gravitational mass, the mass appearing in Newton's law of gravitation: $\mathbf{F} = -\left(Gm_G M_G/r^2\right)\hat{\mathbf{r}}$ (Equation (1.11)). Tower of Pisa experiment shows, using $GM_G^{(E)}m_G/R_E^2 = m_I a_1$ (Newton's second law, Equation (1.12)), that $m_I \propto m_G$.

Section 1.6. An inertial frame is one in which Newton's first law applies (resort to successive approximation to avoid argument being circular).

Section 1.7. Attempts to use Newton's laws in non-inertial frames with constant acceleration require the introduction of artificial 'inertial' forces, $\mathbf{F}_I = -m\mathbf{a}$ (Equation (1.16)), where $\mathbf{a}$ is the frame acceleration.

1.11 Problems

1.1 Show that for N bodies the number of equations of the form of Equation (1.7) that can be written is $N(N-1)/2$. Show that when they are added together they may be cast in the form

$$(N-1)\ (m_1\mathbf{v}_1 + m_2\mathbf{v}_2 + \cdots + m_N\mathbf{v}_N) = \mathbf{constant}\ .$$

Hence deduce Equation (1.8).

1.2 Imagine a very large free-fall lift cabin of typical dimension $r \sim R_E$, where R_E is the radius of the Earth. How might gravity be detectable in such a frame?

The answers to the problems in all chapters can be found in the companion website at www.wiley.com/go/mccall.

2

One-dimensional Motion

2.1 Rationale for One-dimensional Analysis

As we have seen, Newton's laws have quite deep philosophical implications, some of which were discussed in the last chapter. Ultimately, however, the real business of mechanics is to give a mathematical description of motion in response to forces, so in this chapter we begin to examine the simplest cases. It is worth noting the shift here in the role of Newton's laws. Let us make some general comments on the use of Newton's laws to solve mechanics problems:

1. As noted in Chapter 1, for constant mass systems, the first law is really a special case of the second, describing 'force-free' motion.
2. The third law is *only* used in identifying the forces acting, and not in the calculation of motions themselves.
3. Once the forces acting on the various bodies in a problem have been identified, the second law, $\mathbf{F} = m\mathbf{a}$, embodies the *entirety* of the mathematical problem to be solved.

Pause now to consider the potential complexity of the mathematical problem. Forces acting on bodies may in principle depend on position (e.g. gravity, electrostatics), velocity (e.g. friction), or time (e.g. a motor supplying a periodic force). In all generality therefore, we may write for Newton's second law applied to a single body:

$$\mathbf{F} = m\mathbf{a} \text{ or } \begin{aligned} F_x(x, y, z, v_x, v_y, v_z, t) &= m\frac{d^2x}{dt^2}, \\ F_y(x, y, z, v_x, v_y, v_z, t) &= m\frac{d^2y}{dt^2}, \\ F_z(x, y, z, v_x, v_y, v_z, t) &= m\frac{d^2z}{dt^2}. \end{aligned} \tag{2.1}$$

The three equations on the right are the Cartesian components of Newton's second law (Equation (1.4)), expressed in vector form on the left. We distinguish between explicit time dependence and implicit time dependence. A force may be time-dependent as a result of the position of a body changing under the influence of the force. In the new position, the force may take a different value, and since the body's position is time-dependent, then so is the force. This is **implicit time dependence**, so-called because it arises via the position dependence. **Explicit time dependence**, on the other hand, means that even for a stationary body, the force depends on time. Thus if I see a cork bobbing up and down on the surface of

Classical Mechanics – Second Edition From Newton to Einstein: A Modern Introduction Martin W. McCall

a pond, and I wish to keep it still with respect to the bank, then I must supply an explicitly time-dependent force.

There are thus three component equations of motion for the application of Newton's second law to a body. If there are N such bodies, then there are $3N$ equations of motion in all, and matters quickly get out of hand. So we seek to simplify. In this chapter we will simplify by pretending that the world consists of one spatial dimension, and that only one body is subjected to a force.

There is more to it, however, than simple mathematical expediency. Physicists always strive, at least initially, to examine the simplest situations so as to expose the most important features of the phenomenon under scrutiny. By examining Newtonian mechanics in one dimension, we are led to define several naturally occurring quantities that assist our description (e.g. work, energy and power). In one dimension these are, almost of necessity, scalar quantities since in one dimension a 'vector' consists simply of a number together with a sign to indicate direction. However, it turns out that the scalar quantities that we define *remain* as scalar quantities when we move on to discuss motion in higher dimensions.

The other reason for studying one-dimensional motion concerns the physicist's trick of simplifying by viewing a given situation from a different, easier perspective. Newton's laws may be applied in all inertial frames, but the trajectories will look different from one inertial frame to the next. We can often use this latitude of choice of frame to our advantage in problems. Many situations in the real world of three spatial dimensions can be *reduced* to motion in one dimension. Throwing an object at some angle to the ground results in a parabolic (i.e. 2-D) trajectory. However, the horizontal velocity is constant, and so viewed from a frame of reference which moves with this velocity the motion is purely vertical (i.e. 1-D). By studying one-dimensional motion we are actually developing the tools to solve many higher dimensional problems which can be mapped back to one-dimensional motion. When tackling a mechanics problem, the first question to consider is what is the simplest frame in which to view the situation?

2.2 The Concept of a Particle

The fact that we are considering one-dimensional motion means that the location of any body, and more particularly the location through which any force acting on the body passes, is precisely defined. We simply say that the force acts at the point where the body is located. However, since we will want to apply our ideas to more complex situations, it is worth introducing the idealisation of a 'particle'. A particle is the idealisation that an extended body's mass is concentrated at a particular point. To what extent it is appropriate to regard an extended body as a 'particle' depends on the precision required by the problem at hand. Thus, for analysing the trajectory of a cricket ball as it soars beyond the boundary, the particle approximation is adequate, but for analysing the movement off the wicket induced by a bowler's leg break, the ball's 'extended' motion, or spin, must be included and the particle approximation is inappropriate. A particle has no size, structure or orientation.

It turns out, however, that in many cases actual extended bodies behave like particles located at their centre of mass, to be defined later.

2.3 Motion with a Constant Force

Let us assume that a force acts on a particle which does not depend on where the body is situated, how fast it moves, or on time. In one dimension, Newton's second law is

$$F = m\frac{d^2x}{dt^2} \ . \tag{2.2}$$

Since both m and F are constant, so is $d^2x/dt^2 \equiv a$, so that integrating twice with respect to time we obtain for the velocity and position of the particle

$$v = v_0 + at \ , \tag{2.3}$$

$$x = x_0 + v_0 t + \frac{1}{2}at^2 \ , \tag{2.4}$$

where x_0 and v_0 are the particle's initial position and velocity respectively. Eliminating time, t:

$$v^2 = v_0^2 + 2a\,(x - x_0)\,. \tag{2.5}$$

Equations (2.3)–(2.5) are generally the kinematic equations for uniform acceration we first encounter in physics at school, and we sometimes get the impression that they are true in more general situations than they actually are. Let us emphasise, Equations (2.3)–(2.5) apply to uniformly accelerated motion resulting from the action of a constant force acting in one dimension. Incidentally, there's not much difficulty in generalising Equations (2.3) and (2.4) to motion in three dimensions when the body experiences a constant vector acceleration $\mathbf{a}$:

$$\mathbf{v} = \mathbf{v}_0 + \mathbf{a}t \ , \tag{2.6}$$

$$\mathbf{x} = \mathbf{x}_0 + \mathbf{v}_0 t + \frac{1}{2}\mathbf{a}t^2 \ , \tag{2.7}$$

where $\mathbf{x}$ and $\mathbf{v}$ are the vector position and velocity of the particle, with $\mathbf{x}_0$ and $\mathbf{v}_0$ being their initial values.

2.4 Work and Energy

Now integrate Equation (2.2) with respect to x:

$$\int_{x_1}^{x_2} F dx = m\int_{x_1}^{x_2} \frac{dv}{dt}dx = m\int_{v_1}^{v_2} \frac{dx}{dt}dv = m\int_{v_1}^{v_2} v dv \tag{2.8}$$

where we have used $F = mdv/dt$. If F does not depend on position it can be taken outside the integral, obtaining for this special case

$$F \times (x_2 - x_1) = \frac{1}{2}mv_2^2 - \frac{1}{2}mv_1^2 \ . \tag{2.9}$$

Each side of this equation defines a new quantity:

- $F \times (x_2 - x_1)$ is the **work**, W, done *by* a constant force *on* a particle in moving it from x_1 to x_2.

- $K = \frac{1}{2}mv^2$ is the **kinetic energy** (or energy due to motion) of a body of mass m, moving with speed v.

or in terms of momentum, p, the kinetic energy can be expressed as $K = p^2/2m$. What we have found is that

Work done by the (net) force = change in kinetic energy

or, symbolically

$$W_{\text{net}} = \Delta K. \tag{2.10}$$

This is known as the **work–energy theorem** and is again a general result, applicable to position- (but not velocity- or time-) dependent forces.

The SI unit of both kinetic energy and work is the joule which is the work done by a force of 1 N acting over a distance of 1 m (= kinetic energy gained by particle on which such a force acts).

Per se we haven't really learned anything by defining the mechanical quantities work and kinetic energy, they are merely names. But in fact we have completed our first piece of analysis on Newton's second law, through which we have derived the work–energy theorem. We now have an *insight* into mechanical systems which was not obvious from the beginning. If we pick a body in a mechanical system at time 0 and calculate its kinetic energy ($= mv_0^2/2$), and then calculate the same quantity again at time t ($= mv^2/2$), then the difference can be traced to the action of the net force acting over the distance the body moves in that time ($= F \times (x - x_0)$). Relating the motion of bodies to the resultant forces acting, is, after all, the essence of mechanics.

Work thus defined has the same connection with the colloquial notion of work, the exertion of forces. But consider standing still holding out a brick: although we supply a force, the brick does not move, as the force balances that of gravity, so that despite our efforts, no work is apparently done. However, the equilibrium is not static since our muscle fibres are continuously contracting and expanding. The muscles are thus indeed working in the technical sense as the forces they exert move over the short distances of their expansion and contraction, keeping the brick still.

By way of example of the use of the work–energy theorem for a constant force system, consider a particle in freefall under gravity. Gravity provides an approximately constant force close to Earth's surface, so that, taking the z direction as vertically upwards, the z component of the force acting on the particle is

$$F_z = -mg. \tag{2.11}$$

The work done *by* gravity is $W = \int_{z_0}^{z} F_z dz = \int_{z_0}^{z} (-mg)dz = -mg(z - z_0)$. But by the work–energy theorem, $W = -mgz + mgz_0 = \left(mv^2 - mv_0^2\right)/2$, so that

$$\frac{mv^2}{2} + mgz = \frac{mv_0^2}{2} + mgz_0. \tag{2.12}$$

The quantity mgz is referred to as the **potential energy** and charts the work done that would have to be done *against* gravity in raising a body to a particular height, z. This, together with the kinetic energy of a body, constitutes the **total mechanical energy**. We have shown, in this example at least, that the total mechanical energy is the same at some general point of the motion as at the beginning, and is thus a constant in time. Conservation of mechanical energy is a general feature of time and velocity-independent force systems, as we shall see shortly.

2.5 Impulse and Power

Now, writing $a = dv/dt$, integrate Equation (2.2) with respect to time:

$$\int_0^t m\frac{dv}{dt}dt = \int_0^t Fdt. \tag{2.13}$$

Hence

$$\int_{v_0}^{v} mdv = mv - mv_0 = \int_0^t Fdt. \tag{2.14}$$

It is useful to give the right-hand side of Equation (2.14) a name:

- **Impulse** = $\int_0^t Fdt$.

The above shows that the impulse of a force on a body is the change in momentum experienced by the body. Impulse is usually used in contexts where a force acts for a short time, for example when a ball is struck by a bat. The interaction time in such cases is so short that the force may be considered constant over that period, whence impulse $= Ft$.

Another quantity sometimes introduced is the rate of doing work:

- **Power**, P, is the rate of doing work:

$$P = \frac{dW}{dt} = \frac{Fdx}{dt} = F\frac{dx}{dt} = \text{Force} \times \text{velocity}\ . \tag{2.15}$$

The SI unit of power is the watt, equivalent to the rate of working at 1 joule per second. Power is sometimes used in contexts where no net force is acting, which can be a little confusing. When we drive a car along the road with the accelerator depressed it moves because a constant force is applied to the wheels through the burning of fuel. However, resistance from the road and the rotary motion of the car's transmission opposes the car's motion so that in the steady state no *net* force is applied. According to Newton's first law the car then moves with constant velocity, and the power developed by the engine is that required to overcome friction and to maintain the car's speed. Perhaps the right pedal would be better named the 'velocirater'!

2.6 Motion with a Position-dependent Force

Although we've worked it out for a constant force, the definition $W = \int_{x_0}^{x} Fdx$ covers the case $F = F(x)$ as well. Many forces depend on the location of a body and not explicitly on time or the body's velocity. We refer in such cases to a **field of force**. Gravity is the most familiar example, though we have to go to a height that is comparable with the Earth's radius before the acceleration due to gravity changes noticeably. In the context of one-dimensional motion, the position dependence is with respect to a single spatial variable, which we call x. Unlike the case of position-independent forces, $m\ddot{x} = F(x)$ can no longer be integrated with respect to time directly (since $x = x(t)$), so we must proceed differently. First, let us generalise the definition of work from Equation (2.9):

- The **work**, W, done by a position-dependent force, $F(x)$, is given by

$$W = \int_{x_0}^{x} F(x)dx \quad . \tag{2.16}$$

Note that this includes the old definition as a special case. Everything can now proceed as before up to and including the work–energy theorem, so that now

$$W = \int_{x_0}^{x} F(x)dx = \frac{1}{2}mv^2 - \frac{1}{2}mv_0^2 = \Delta K. \tag{2.17}$$

As defined, the work done by the force clearly depends only on the starting point x_0 and the final point x. This dependence can be emphasised by writing $W = -U(x) + U(x_0)$, the so-called **potential energy function** being defined as

$$U(x) = -\int F(x)dx, \text{ or } F(x) = -\frac{dU}{dx} \; . \tag{2.18}$$

The choice of sign is simply a matter of convention. The work–energy theorem in symbols becomes $-U(x_0) + U(x) = K - K_0$, or

$$K + U(x) = K_0 + U(x_0) = \text{ constant } = E \; . \tag{2.19}$$

We have thus generalised the concept of mechanical energy, E, so that it now includes position-dependent forces, and we have found, as a direct consequence of Newton's second law, that in such cases it is a conserved quantity. We have developed the work–energy theorem slightly in that by writing the work done as the difference between a potential energy function, $U(x)$, evaluated at two points along the motion, a conserved quantity emerges, being the sum of the kinetic energy and the potential energy. Note also that $U(x)$ is only defined up to a constant, in that adding a constant to both sides of Equation (2.19) does not alter its content, and the work–energy theorem is unchanged. From the definition, $U(x)$ is seen to be the work done in moving a particle *against* the force field, $F(x)$. If the particle is then released so that the force $F(x)$ moves the particle back to its original position, x_0, then a positive amount of work is done. This work is, by the work–energy theorem, equal to the increase in the particle's kinetic energy, but is also numerically equal to $U(x) - U(x_0)$. So the function $U(x)$ is seen to measure the potential to recover the work done against a force as kinetic energy, hence its designation as potential energy.

Let us just check that the mechanical energy is constant, by differentiating it with respect to time:

$$\frac{dE}{dt} = \frac{dK}{dt} + \frac{dU}{dt} \; ,$$

$$= \frac{d\left(\frac{1}{2}mv^2\right)}{dt} + \frac{dU(x)}{dt} = mv\frac{dv}{dt} + \frac{dU}{dx}\frac{dx}{dt},$$
$$= vF - Fv = 0\,,$$

as required. Note that this only works because F (and therefore U) has no *explicit* time dependence. For time-dependent forces, the mechanical energy is *not* conserved. For this reason, position-dependent forces are sometimes referred to as **conservative forces**. We have worked solely in one dimension for simplicity, although the concept of a conservative force carries over to higher dimensions – see Chapter 7.

By way of example, let us consider the natural generalisation of the body falling under gravity considered previously. Consider the following problem:

A body is released from rest a long way from the Earth. With what speed will it hit the Earth?

Applying the work–energy theorem, $mv^2/2 - mv_0^2/2 = \int_{z_0}^{z} F(z)dz$, where the initial velocity, $v_0 = 0$, and $F(z) = -GMm/z^2$ is the gravitational force at distance z from the centre of the Earth. (M is the mass of the Earth $= 5.98 \times 10^{24}$ kg, m is the mass of the body and G is the universal gravitation constant $= 6.672 \times 10^{-11}\ \mathrm{N\,m^2\,kg^{-2}}$). Hence

$$\frac{1}{2}mv^2 = \int_{z_0}^{z}\left(-\frac{GMm}{z^2}\right)dz = \left.\frac{GMm}{z}\right|_{z_0}^{z} = -U(z) + U(z_0). \tag{2.20}$$

Here, $U(z) = -GMm/z$, up to an arbitrary constant. If the particle starts a long way from the Earth, then $z_0 \to \infty$. The final position is $z = R_e$ where R_e is the Earth's radius ($= 6.37 \times 10^6$ m), so finally

$$\frac{1}{2}mv^2 = \frac{GMm}{R_e}, \text{ or } v = \sqrt{\frac{2GM}{R_e}}\,. \tag{2.21}$$

Putting in the parameters, $v = 1.12 \times 10^4\ \mathrm{m\,s^{-1}}$, or $v \approx 11\ \mathrm{km\,s^{-1}}$. The argument works the other way round, that is $v \approx 11\ \mathrm{km\,s^{-1}}$ is the minimum initial speed which must be imparted to a body at the Earth's surface so as to escape Earth gravity, that is its **escape velocity**. The escape velocity is a characteristic velocity of a gravitating body.

2.7 The Nature of Energy

We have seen that for conservative forces, a potential energy function can be defined which, when added to a body's kinetic energy, results in a conserved quantity, the total mechanical energy, E. Let us emphasise the origin of terms in the conservation of energy, by writing it as

$$E \equiv \frac{1}{2}mv^2 - \int_{x'}^{x} F(x)dx = \frac{1}{2}mv_0^2 - \int_{x'}^{x_0} F(x)dx \tag{2.22}$$

where x' is an arbitrary position. Because of this arbitrariness, any constant can be added to E without changing the physical content of energy conservation. Once the arbitrary constant is fixed, the resultant mechanical energy is conserved. We emphasise at this stage that this is merely a deduction from Newton's second law applied to position-dependent forces. For velocity-dependent forces, such as friction, E is no longer conserved. The question then naturally arises: can the definition be usefully extended so that energy conservation is true even when nonconservative forces are acting? For friction, we note that the bodies rubbing together get hot, and we can then define 'heat energy' as being equal to the product of the mass of the body, its specific heat capacity and its rise in temperature. This is found to account for most of the frictional dissipation. When there are still minor discrepancies, other forms of energy, such as electromagnetic, nuclear, elastic, etc. can be introduced. When we are done we hope to find that the total energy is constant.

The procedure of *how* to define the new forms of energy is quite subtle, and again starts to look like a circular contrivance. Moreover, keeping track of all the different forms of energy as they undergo their various transmutations is practically impossible, so the conservation of energy would seem to have vacuous content.

This all seems somewhat unsatisfactory, as we are accustomed to thinking of the conservation of energy as having a stronger status, and the process of inventing new forms of energy simply to preserve its conservation seems a little arbitrary. Special relativity removes much of this arbitrariness, however. There, the interrelationships between mass, momentum and energy are such that the bookkeeping of all conceivable types of energy is accounted for. The arbitrary constant in the above definition of mechanical energy is also removed. At this stage, therefore, we must consider the conservation of mechanical energy to be an important deduction from Newton's second law, and await its clarification in relativity.

At the deepest level, however, energy remains elusive. Feynman, in his famous Lectures on Physics,[1] drew a beautiful analogy to illuminate the nature of energy. Imagine a child with some blocks, who, in the course of playing with them, hides some in a bucket, throws some in the bath, puts some under the carpet and retains a few with which to build. His ingenious mother, who wants to check that no blocks are lost, counts those she can see, and adds the number whose existence she can *infer* by weighing the bucket, measuring the rise in the water level, seeing the humps in the carpet etc. Whatever the child does, the mother knows that the number of blocks is fixed. Now the punchline: *for energy there is nothing equivalent to the visible blocks remaining*! Energy is entirely inferential, through the 'traces' it leaves as it is transmuted from one form to another. We can count it, but it is not 'there' in any physical sense. It is truly elusive.

2.8 Potential Functions

So far we have presumed that a body is moving under the influence of a conservative force and deduced that its total mechanical energy, potential plus kinetic, is a constant. Now we turn the argument around. We presume that we are given a particular potential function, $U(x)$, from which the force field may be derived as $F = -dU/dx$, and try to deduce the principal features of the motion for different total energies, E. For these purposes, then, we write

$$K = E - U \,. \tag{2.23}$$

Since $K = mv^2/2$ is definitely positive, then $E \geq U$. A simple example is the potential function $U = k/x$ for $x > 0$, which is plotted for various values of k in Figure 2.1. Since

[1] 'Feynman, R.P., Leighton, R.B. and Sands, M. (1963) The Feynman Lectures on Physics, Vol. 1, Addison-Wesley.

$F = -dU/dx = k/x^2$, the potential function clearly represents an inverse square force law. If k is negative, then the force acts towards the origin and is thus *attractive* to that point. This is the case for gravitation where $k = -GmM$, and the mass m is attracted towards the mass M. For the electrostatic interaction between two charges, q_1 and q_2, $k = q_1 q_2/4\pi\varepsilon_0$, which may be of either sign. For like charges ($k > 0$) the force is *repulsive*, whilst for unlike charges ($k < 0$) the force is *attractive*.

You can think of the potential function $U(x)$ as the profile of a hill on which is placed a ball. The ball runs down the hill, that is in the direction of $(-dU/dx)$. This is the motivation for the minus sign in the definition $U = -\int F dx$.

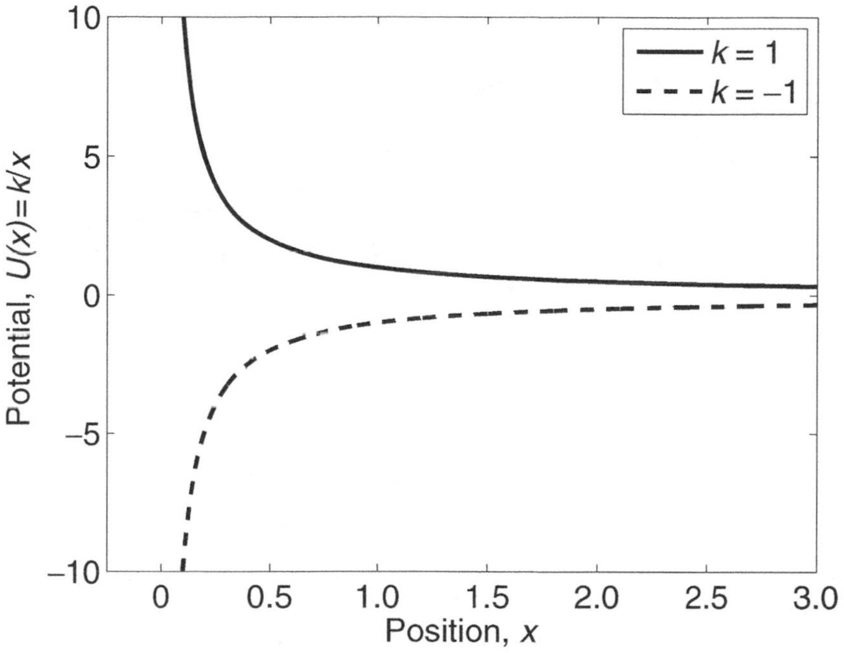

Figure 2.1 k/x potential functions.

Now consider the more complicated potential function shown in Figure 2.2. The total energy $E = K + U$ is conserved, so that since K is positive, various motions may be deduced for different values of E:

1. If $E = E_F$ then for $x \geq x_0$ the particle is **free** and rolls down the potential hill to $x = +\infty$. Note that $x < x_0$ is not possible as this violates the condition that K is positive.
2. If $E = E_B$ and $x > x_3$ then the particle is free. If, however, x lies between x_1 and x_2 initially, then it must remain between these limits forever, as it has insufficient energy to escape out of the potential well. The particle is said to be **bound**. At either x_1 or

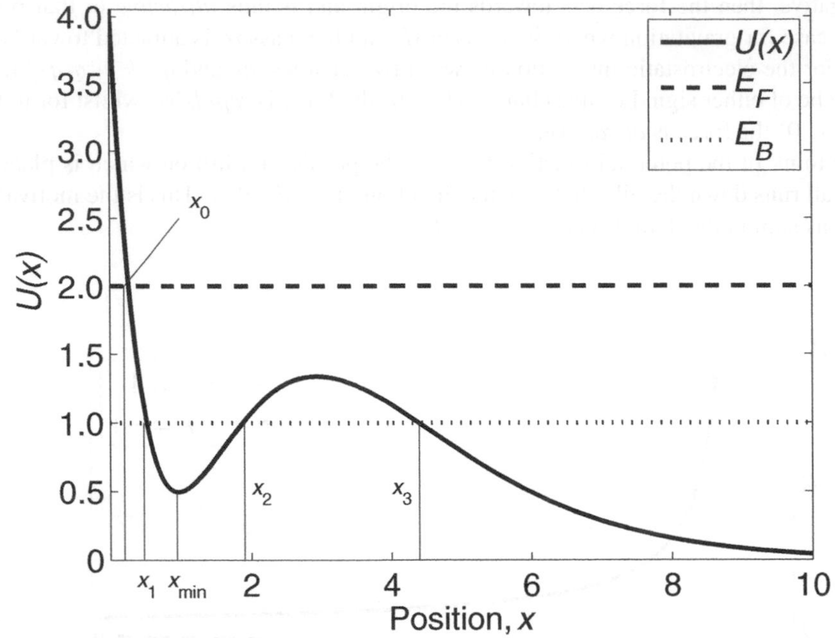

Figure 2.2 A more complicated potential function.

x_2, we have $K = 0$ and the particle is momentarily at rest, whilst at $x_{\min}$, we have $F = -dU/dx = 0$ and the particle is momentarily in equilibrium.

3. The regions for which K is negative, that is $x < x_1$ or $x_2 < x < x_3$, are **forbidden**.

We will analyse the oscillatory motion within the potential well in the next section. The above reasoning is classical and based on the tenets of Newton's laws. One of the most surprising features of quantum mechanics is that small violations of the above reasoning are allowed. 'Small' here means that the closer the constant energy line approaches the top of a potential hill from below, the more probability there is that a particle will 'tunnel' from one classical region ($x_1 < x < x_2$) to another ($x > x_3$), that is it becomes possible for a particle to 'jump the gap' across the forbidden region ($x_2 < x < x_3$). Such a process lies at the heart of radioactive decay, which classically would not occur.

The type of classical reasoning employed here to determine the qualitative feature of a motion (i.e. whether it is bound or unbound) is quite interesting. Ironically qualitative analysis, with its overtones of Aristotelian imprecision, has experienced something of a revival in the last few decades. For longer than that it has been recognised that most problems in mechanics are analytically intractable. But in the modern branches of mechanics known as nonlinear dynamics and chaos, for example, it is now appreciated that the extent to which important information can be gleaned from analysing potential-like functions. Whether a body escapes to infinity, whether a given system (e.g. the solar system) is ultimately stable, or what kinds of periodic behaviour can occur, are all questions concerning the *general*, or global, behaviour of systems, which can often be answered without solving for detailed particle trajectories. We will find in Chapter 7 that the ideas of qualitative analysis using potential functions are invaluable in analysing gravitational orbits.

2.9 Equilibria

What are the requirements for a particle to be in equilibrium? The colloquial notion of 'equilibrium' is that of a state of balance, or of rest. However, a balance of forces is *necessary*, but not *sufficient* for a particle to be at rest, since, by Newton's first law, when there is no net force on the particle, it may move with constant velocity. Conversely, a particle may be momentarily at rest, while the balance of forces may not vanish. A moment later the particle will have moved due to the net force. Mechanically, 'equilibrium' and 'at rest' should not be identified. Equilibrium is defined as there being no net force on the particle. For the potential functions introduced above we see that the equilibrium condition is that $F = -dU/dx = 0$. Thus a turning point in the potential function, at say $x = a$, corresponds to an equilibrium point. Now *if* a particle is placed at rest at $x = a$, it will remain at rest forever. However, how the particle behaves if it is displaced slightly will depend on the curvature of the potential function at that point. There are therefore different sorts of equilibria, depending on the value of the second derivative of the potential function at $x = a$ (see Figure 2.3). The three cases are:

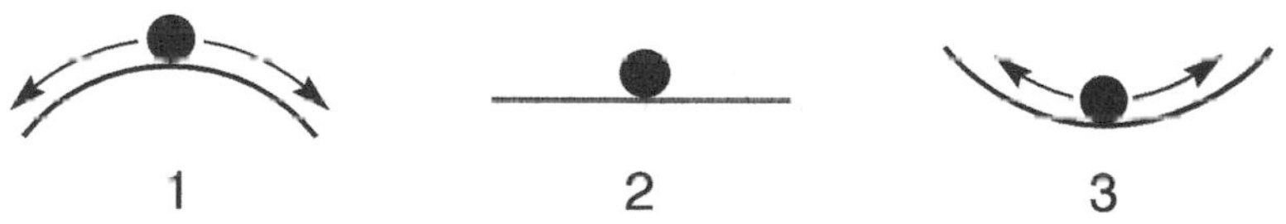

Figure 2.3 Equilibrium types: (1) unstable, (2) neutral, (3) stable.

1. $\left.\frac{d^2U}{dx^2}\right|_{x=a} = U''(a) < 0$ When displaced from equilibrium the ball will run away from $x = a$, that is the equilibrium is **unstable**.
2. $\left.\frac{d^2U}{dx^2}\right|_{x=a} = U''(a) = 0$ Small displacements from equilibrium result in no force on the ball. The equilibrium is **indeterminate** or locally **neutral**. The stability of such equilibria must be determined by looking at higher derivatives of the potential function.
3. $\left.\frac{d^2U}{dx^2}\right|_{x=a} = U''(a) > 0$ Here, a small displacement results in an attractive force back towards the equilibrium point, so the ball always stays in the vicinity of $x = a$. The equilibrium is **stable**.

2.10 Motion Close to a Stable Equilibrium

The case of stable equilibrium is worth looking at further, because it is possible to deduce general properties of the motion confined in a potential well. When a particle is displaced from a stable equilibrium point it experiences a restoring force always directed back towards the equilibrium point. The best way to see this is to approximate the potential function near the equilibrium point as a Taylor expansion:

$$U(x) = U(a) + U'(a)[x-a] + \tfrac{1}{2}U''(a)[x-a]^2 + \cdots . \tag{2.24}$$

Thus, introducing the variable $X \equiv x - a$, which measures the displacement from equilibrium, the force is given by

$$F = -\frac{dU}{dX} \approx -U''(X=0)[X] \tag{2.25}$$

where the approximate equality arises because the higher-order terms are now being dropped.[2] Since $U''(X=0)$ is a positive constant, a positive displacement from equilibrium results in a negative force. Similarly, for a negative displacement the force is positive. Thus the force is of the form of a spring obeying Hooke's law, where the restoring force is proportional to the displacement, the spring constant $k = U''(a)$. The equation of motion for the system is, by Newton's second law,

$$m\ddot{X} = -kX. \tag{2.26}$$

The resultant motion is simple harmonic, which will be analysed in detail in the next chapter. For now let us simply note that the angular frequency of the motion and the period of the oscillations are given respectively by

$$\omega = \sqrt{\frac{k}{m}} = \sqrt{\frac{U''(X=0)}{m}} \tag{2.27}$$

and

$$T = \frac{2\pi}{\omega} = 2\pi\sqrt{\frac{m}{U''(x=a)}}. \tag{2.28}$$

The results of this section have rather wide applicability. Almost *any* system observed to be in equilibrium – a pendulum at the bottom of its swing, atoms bound together in a molecular or crystal structure, a guitar string – all are necessarily in *stable* equilibrium (otherwise the inevitable presence of small perturbations would drive the system away from equilibrium). Since most observed equilibria are stable, the above analysis can be used to determine what will happen to the system when it is slightly perturbed. Thus Hooke's 'law' is *not* a fundamental law of physics – it is simply the consequence of analysing systems near a stable equilibrium point. All that we have said applies to systems for which the force is conservative, so that a potential energy function is defined. It turns out that even in the presence of small amounts of dissipation, the frequency of the oscillations near equilibrium is only slightly modified (see Chapter 3) so that the results are quite general.

Let us now turn to the discussion of unstable equilibria.

2.11 The Stability of the Universe

We are now going to solve a problem of literally cosmic significance that eluded even Newton. Consider two identical bodies of mass M, fixed at $x = \pm D$. The potential function

[2] Strictly speaking when we change the variable according to $X \equiv x - x_0$, we are defining a *new* function $\tilde{U}(X) = U(x - x_0)$.

experienced by a third identical mass, which is free to move under the gravitational influence of the other two masses, is given by

$$U_1(x) = -\frac{GM^2}{|x-D|} - \frac{GM^2}{|x+D|} \,. \tag{2.29}$$

A plot of this function is shown in Figure 2.4. The function has singularities at $x = \pm D$ and

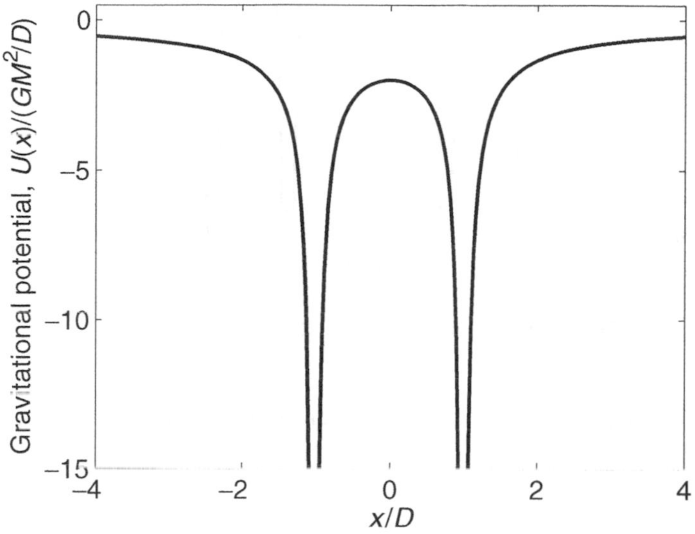

Figure 2.4 Potential function of Equation (2.29).

an equilibrium point at $x = 0$, where $U_1(0) = -2GM^2/D$. By the considerations of Section 2.9 we can see that the equilibrium point is unstable. This can either be inferred from the shape of the graph at $x = 0$, or by calculating $d^2U_1(0)/dx^2$ explicitly as

$$\left.\frac{d^2U_1(0)}{dx^2}\right|_{x=0} = -\frac{4GM^2}{D^3} < 0 \,. \tag{2.30}$$

Now fix two further bodies, also of mass M, at $x = \pm 2D$, resulting in the new potential function

$$U_2(x) = -\frac{GM^2}{|x-D|} - \frac{GM^2}{|x+D|} - \frac{GM^2}{|x-2D|} - \frac{GM^2}{|x+2D|}, \tag{2.31}$$

which is sketched in Figure 2.5. Two new singular points appear at $x = \pm 2D$, whilst the presence of the additional masses reduces the potential at the origin to $U_2(0) = -3GM^2/D$. The origin is still an unstable equilibrium point. A small perturbation of a test mass placed at the origin will fall down one of the wells either side. How will this process continue as more

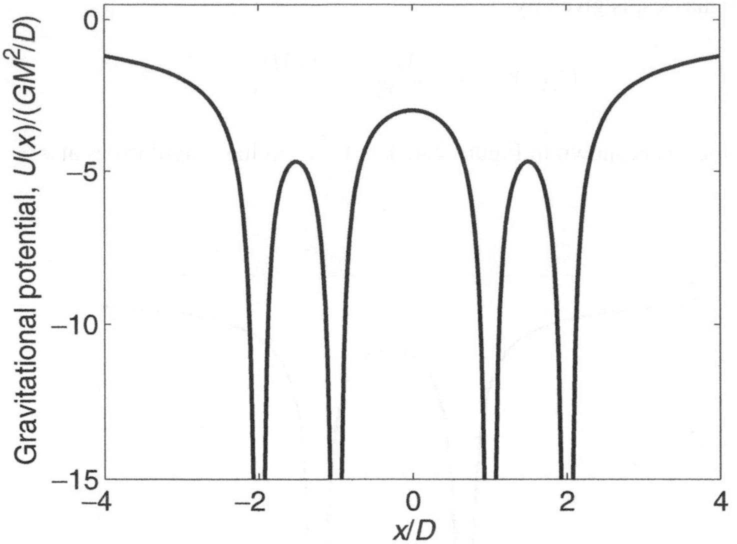

Figure 2.5 Potential function of Equation (2.31).

mass pairs are added? Figure 2.6 shows the potential for seven mass pairs. For N mass pairs the second derivative at the origin is

$$\left.\frac{d^2U_N(0)}{dx^2}\right|_{x=0} = -\frac{4GM^2}{D^3}\sum_{n=1}^{N}\frac{1}{n^3} < 0\,, \tag{2.32}$$

which is still negative, so that the origin remains a point of unstable equilibrium. The value of the potential at the origin is

$$U_N(0) = -\frac{2GM^2}{D}\sum_{n=1}^{N}\frac{1}{n}\,. \tag{2.33}$$

$U_N(x)$ is the potential function felt by a test mass due to the placement of all the other mass pairs at $x = \pm nD$ $(n = 1, 2, \cdots, N)$. The mass pairs themselves will each be immersed in a *different* potential, one that will be asymmetric on account of there being more mass to one side than the other, and will attract the system towards the origin. There is a special case, however, when the potential felt by *any* mass in the line is the same as that of the test mass at the origin, namely when $N \to \infty$ and we have an infinite line of masses.[3] It might be

[3]There is a technicality associated with an infinite line of masses in that as $N \to \infty$, the series $\sum_{1}^{\infty}\frac{1}{n}$ diverges, so that $U_\infty(0)$ is infinitely negative. However, the problem can be overcome by increasing the potential a little bit each time a mass pair is added so as to keep the potential at the origin finite (recall that adding a number does not change the physical content of the potential function). The nature of the stability of the potential function (which is all we are interested in) is determined solely by the second derivative of the potential function, which remains finite.

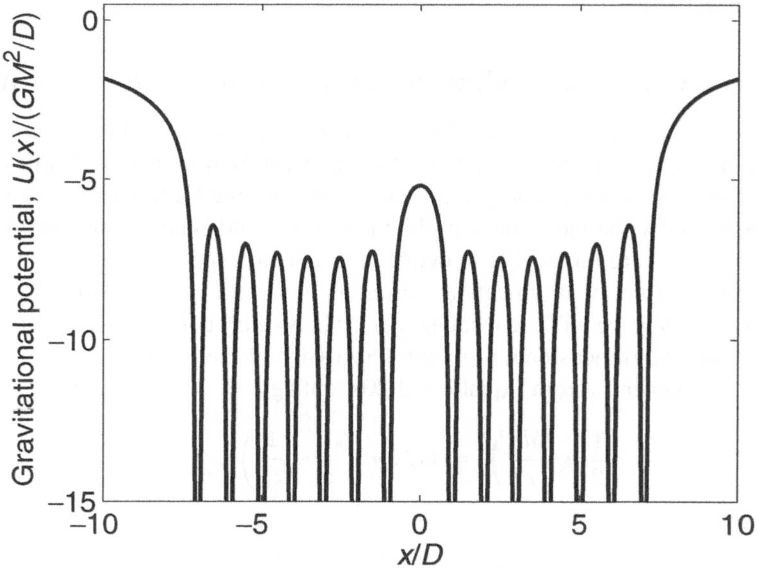

Figure 2.6 Potential function of seven mass pairs.

thought that this configuration is stable as then effectively everything is attracted equally to everything else. We can calculate explicitly the stability of the test mass at the origin as

$$\left.\frac{d^2U_\infty(0)}{dx^2}\right|_{x=0} = -\frac{4GM^2}{D^3}\sum_{n=1}^{\infty}\frac{1}{n^3}\,. \tag{2.34}$$

We need only note that the summation converges to a positive number[4] to see that $d^2U_\infty(0)/dx^2 < 0$, so that the point at the origin is once again an unstable equilibrium. But we also know that this conclusion holds for *any* point in the line, so that a line of infinite masses, now all free to move, is actually gravitationally *unstable*. The slightest perturbation of the masses about their equilibrium positions will cause the masses to be locally attracted to each other and the distribution will never stabilise. Such a line of masses is *dynamic*. Although as a model universe, a line of equally spaced identical masses is very naive, the conclusion holds for more refined models involving higher dimensions and unequal/nonuniform mass distributions. We have therefore shown that even an infinite universe cannot be stable under Newtonian gravitational interactions.[5]

Let us not understate Newton's accomplishment in enunciating the laws of mechanics, which undoubtedly ranks as a defining achievement in scientific history. And yet, what should

---

[4]In fact $\sum_1^\infty \frac{1}{n^p}$ converges for all $p > 1$. For $n = 3$, $\sum_1^\infty \frac{1}{n^3} = 1.202...$, known as Apéry's constant.

[5]A similar, though more sophisticated conclusion, lies at the heart of general relativity. Einstein described his contortions to preserve a static universe within his theory, through the introduction of the so-called cosmological constant, as 'the greatest blunder of my life'.

have been its crowning feat – the deduction of a *dynamic* universe – was only discovered experimentally much later.

2.12 Trajectory of a Body Falling a Large Distance Under Gravity

This section is a bit more advanced but, I think, interesting. At the end of Section 2.6 we used energy arguments to calculate the speed with which a body would hit the Earth's surface when released from rest at a great height z_0, measured from the Earth's centre. Notice that we did not formally solve the mechanical problem, which would entail giving the position as a function of time, but were rather able to deduce the collision speed using the conservation of mechanical energy. Energy arguments are often able to yield useful information without solving the problem. Now we will see what happens when we try to find the body's trajectory, $z(t)$, for this strictly one-dimensional problem. Discussion of the gravitational orbits in 3-D will be given in in Chapter 7. From Equation (2.20), setting $v = dz/dt$ we have

$$\frac{1}{2}m\left(\frac{dz}{dt}\right)^2 = GMm\left(\frac{1}{z} - \frac{1}{z_0}\right) . \tag{2.35}$$

Rearranging yields

$$\frac{dz}{dt} = -(2GM)^{1/2}\left(\frac{1}{z} - \frac{1}{z_0}\right)^{1/2} , \tag{2.36}$$

where taking the negative root indicates that the body is *falling*. Finding the trajectory $z(t)$ is then equivalent to solving the integral

$$\int_{z_0}^{z}\left(\frac{1}{z} - \frac{1}{z_0}\right)^{-1/2} dz = -(2GM)^{1/2}\, t .$$

The integral is actually a bit tricky, so you can either look it up, or use the substitution $z = z_0 \sin^2\theta$ to find

$$\frac{\pi}{2} - \arcsin\left(\frac{z}{z_0}\right)^{1/2} - \frac{1}{2}\sin\left[2\arcsin\left(\frac{z}{z_0}\right)^{1/2}\right] = \left(\frac{2GM}{z_0^3}\right)^{1/2} t . \tag{2.37}$$

We wanted to find $z(t)$, but Equation (2.37) actually provides $t(z)$. We cannot invert Equation (2.37) to find $z(t)$ (try it!), but we can at least check that $z = z_0$ when $t = 0$, and that differentiating Equation (2.37) with respect to time brings us back to Equation (2.36) (to check the latter you will require $d\arcsin(x)/dx = (1 - x^2)^{-1/2}$). If the attracting mass is concentrated at the origin, then, by setting $z = 0$, Equation (2.37) can be used to calculate the fall time as

$$\tau = \frac{\pi}{2}\left(\frac{z_0^3}{2GM}\right)^{1/2} . \tag{2.38}$$

By dividing the region from $z = 0$ to $z = z_0$ up into equally spaced increments, we can calculate the corresponding values of t. The result is a numerical determination of the trajectory $z(t)$ and is plotted in Figure 2.7. Surprisingly, I have not seen this solution given in other textbooks. If you are interested in exploring the rather curious features of this solution, see 'Gravitational Orbits in One Dimension', *American Journal of Physics*, **74**, 1115–1119, (2006). You can also see if Milton was correct in his estimate of nine days for the angels to fall from heaven given in his epic poem:

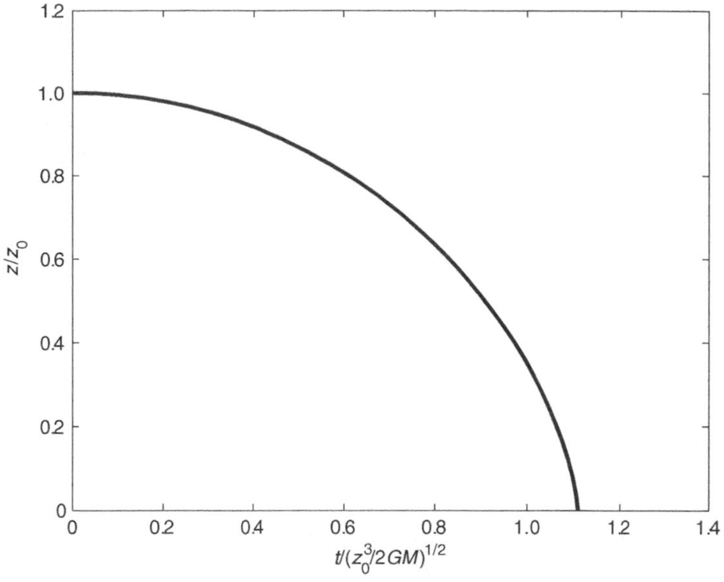

Figure 2.7 Trajectory $z(t)$ of particle falling from a great height, determined from Equation (2.37).

And Chrystall wall of Heav'n, which op'ning wide,
Rowld inward, and a spacious Gap disclos'd
Into the wastful Deep; the monstrous sight
Strook them with horror backward, but far worse
Urg'd them behind; headlong themselvs they threw
Down from the verge of Heav'n, Eternal wrauth
Burnt after them to the bottomless pit.
Hell heard th' unsufferable noise, Hell saw
Heav'n ruining from Heav'n, and would have fled
Affrighted; but strict Fate had cast too deep
Her dark foundations, and too fast had bound.
Nine dayes they fell; confounded Chaos roard,
And felt tenfold confusion in thir fall
Through his wilde Anarchie, so huge a rout
Incumberd him with ruin: Hell at last
Yawning receavd them whole, and on them clos'd,

Paradise Lost

2.13 Motion with a Velocity-dependent Force

To complete our study of one-dimensional motion we now consider a nonconservative force, i.e. friction. Now the force is presumed to depend on velocity, $F = F(v)$. It is no longer possible to define a potential function from which the force may be derived, so it is called a **nonconservative** force. The simplest possible assumption is that $F \propto v$ which corresponds to fluid friction at low speeds. For higher speeds it turns out that frequently $F \propto v^2$ and is referred to as **drag**. Thus for a model system, consider the forces acting on the parachutist shown in Figure 2.8. Taking downwards as the positive direction, the velocity-dependent

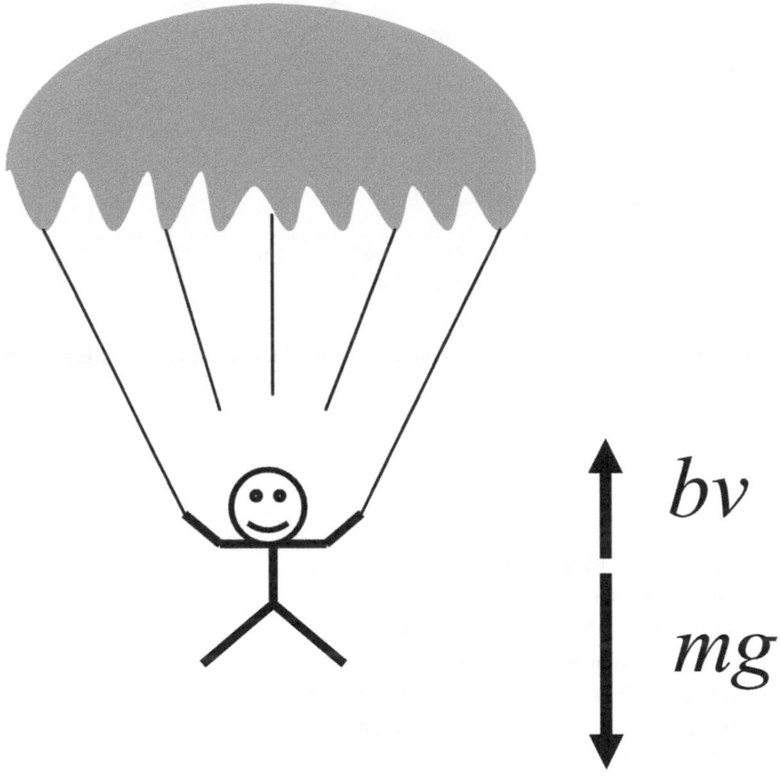

Figure 2.8 Parachutist experiencing simple linear resistance.

resistive force is given by

$$F_{\text{resistance}} = -bv \;, \tag{2.39}$$

where the coefficient b, with units kg s^{-1}, characterises the viscoscity of the fluid in which the body is moving, in this case air. Note that the resistive force opposes the motion. Presuming that the parachutist is not too far from the Earth's surface the gravitational force is constant and equal to mg and so by Newton's second law the equation of motion is given by

$$mg - bv = m\frac{dv}{dt} \;. \tag{2.40}$$

This may be recast as the following integration

$$m\int_0^v \frac{dv}{mg - bv} = \int_0^t dt\,, \tag{2.41}$$

which when carried out yields the velocity as a function of time:

$$v = \frac{mg}{b}\left[1 - \exp\left(-\frac{bt}{m}\right)\right]\,. \tag{2.42}$$

The form of this velocity-dependence is illustrated for various parameter values in Figure 2.9. As $t \to \infty$, $v \to mg/b \equiv v_\infty$, that is the body approaches its so-called **terminal velocity**. The body is then moving with constant speed, so by Newton's first law no force is acting, and gravity is then exactly opposed by the viscous force. The rate at which terminal velocity is approached is governed by the time constant m/b. In terms of the terminal speed, Equation (2.42) may be written as

$$v = v_\infty\left[1 - \exp\left(-\frac{gt}{v_\infty}\right)\right]\,. \tag{2.43}$$

Thus if the system is changed so as to increase the terminal velocity (by increasing the mass

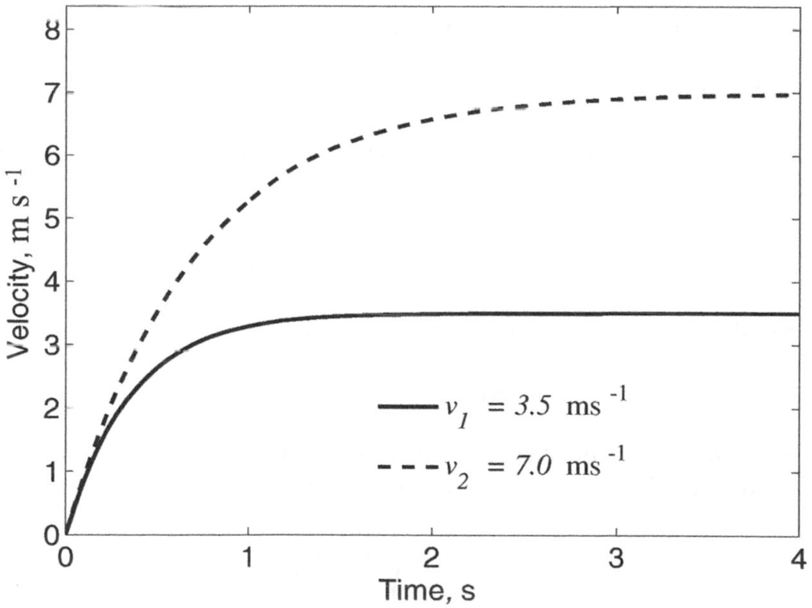

Figure 2.9 Velocity-time graph showing resistive motion for different terminal velocities. Parameters: $b = 140$ kg s^{-1}, $m = 50$ kg (solid line), $m = 100$ kg (broken line).

or decreasing the viscosity), then the time taken to reach terminal velocity is also increased.

If you want a challenge, try figuring it all out for high-speed resistance with a drag force given by $F = -Dv^2$ where D = drag coefficient. This corresponds to the freefall skydiver.[6]

2.14 Summary

Section 2.2. Particles have no size, structure or orientation.

Section 2.3. Newton's second law for motion of a particle experiencing a constant force yields a constant acceleration, $d^2x/dt^2 = F/m = a$ (Equation (2.2)), which when integrated twice gives $v = v_0 + at$, Equation (2.3), and $x = x_0 + v_0 t + \frac{1}{2}at^2$, Equation(2.4).

Section 2.4. Integrating Equation (2.2) with respect to x gives

$$F(x - x_0) = \frac{1}{2}mv^2 - \frac{1}{2}mv_0^2 \tag{2.9}$$

where the left-hand side is the **work done** by the force and the right-hand side is the resultant change in the particle's **kinetic energy**. This generalises for position-dependent forces to

$$W = \int_{x_0}^{x} F(x)dx = \frac{1}{2}mv^2 - \frac{1}{2}mv_0^2 = \Delta K \,, \tag{2.17}$$

the result being known as the **work–energy theorem**.

Section 2.5. **Impulse** = $\int_0^t F dt$. **Power** , $P = \frac{dW}{dt} = Fv$ (Equation (2.15)).

Section 2.6. For position-dependent, or **conservative**, forces it is possible to define a **potential function** according to

$$U = -\int F(x)dx \text{ , or } F = -\frac{dU}{dx} \,, \tag{2.18}$$

in which case the work–energy theorem may be recast as

$$K + U(x) = K_0 + U(x_0) = \text{ constant } = E \,, \tag{2.19}$$

where E is the **total mechanical energy**. For a gravitational field $U = -GmM/r$, and the so-called **escape velocity** from a spherical mass of radius $r = R$ is given by

$$v_{\text{escape}} = \sqrt{\frac{2GM}{R}} \,. \tag{2.21}$$

Section 2.8. The motion of a particle under the influence of a conservative force may be classified as **bound**, **unbound** or **forbidden**, by noting where the particle lies with respect to the (constant) energy line – see Figure 2.2.

[6] Answer: $v(t) = [v_0 + v_\infty \tanh(gt/v_\infty)]\,[1 + (v_0/v_\infty)\tanh(gt/v_\infty)]^{-1}$, where v_0 is the initial velocity, $\tanh(x) = (e^x - e^{-x})/(e^x + e^{-x})$, and $v_\infty = (mg/D)^{1/2}$ is the terminal velocity.

Section 2.9. Near equilibrium the particle's stability is classified according to

$$U''(x=a) = \left.\frac{d^2U}{dx^2}\right|_{x=a} > 0\,(<0) \;\Rightarrow\; \text{stable (unstable) equilibrium}\,.$$

Section 2.10. Near a stable equilibrium point the motion is simple harmonic with period

$$T = \frac{2\pi}{\omega} = 2\pi\sqrt{\frac{m}{U''(x=a)}}\,. \tag{2.28}$$

Section 2.11. An infinite Newtonian universe is inherently unstable.

Section 2.13. Velocity-dependent forces of the form

$$F(v) = -bv \tag{2.39}$$

give an equation of motion

$$mg - bv = m\frac{dv}{dt}\,. \tag{2.40}$$

The solution of this equation is

$$v = \frac{mg}{b}\left[1 - \exp\left(-\frac{bt}{m}\right)\right]\,, \tag{2.42}$$

showing that after $t >> m/b$, the particle reaches **terminal velocity**, $v_\infty \equiv mg/b$.

2.15 Problems

Where required take a value for the acceleration due to gravity: $g = 10\,\mathrm{m\,s^{-2}}$.

2.1 A car travels for 5 km at 100 km h^{-1}, then for 5 km at 25 km h^{-1}. What is its average speed over the 10 km journey?

2.2 An object is released from rest at the top of the Tower of Pisa, 60m high. When it reaches the ground, it rebounds such that its upward speed is half the downward speed at impact. The object is in contact with the ground for 0.1 s. Calculate:

(a) the time taken to fall from the top of the tower to the ground;
(b) the object's speed when it hits the ground;
(c) its average acceleration during the bounce;
(d) the height reached following the rebound.

2.3 The coordinates of a particle (in metres) are given by $x = \cos 2t$ and $y = \sin 2t$. Calculate the magnitude and direction of (i) the particle's velocity, and (ii) its acceleration at $t = 0$, $t = \pi/2$, $t = \pi$ s. Describe the particle's motion (in a few words, not equations).

2.4 A particle's trajectory is described by the position vector: $\mathbf{r} = (3t^2 - 2t)\mathbf{i} - t^3\mathbf{j}$ metres. Calculate:

(a) its speed at $t = 2$ s;
(b) its acceleration at $t = 4$ s;
(c) the average acceleration between $t = 1$ s and $t = 3$ s.

2.5 A particle of mass 1.5 kg has an initial velocity of $(2\mathbf{i} + 3\mathbf{j})$ m s^{-1}. It is acted on by a force of $(4\mathbf{i} - \mathbf{j})$ N for 3 s. What is its final velocity?

2.6 A particle of mass 2 kg is observed to follow a trajectory $\mathbf{r} = \mathbf{i} + 15t\mathbf{j} - 5t^2\mathbf{k}$ where $\mathbf{i}, \mathbf{j}$ and $\mathbf{k}$ are orthogonal unit vectors. Calculate the particle speed at time $t = 2$ s, and the force on the particle. Describe the motion.

2.7 A book, initially at rest, is pushed along a horizontal table top by a horizontal force of 2 N. After 1 s it has been moved 1.5 m. The force of friction is 0.4 N.

(a) How much work is done on the book by the 2 N force?
(b) How much work is done on the book by friction?
(c) What is the book's kinetic energy after 1 s?
(d) What power is required?

2.8 If 10 kW are required to drive a 1000 kg car at 72 km h^{-1} on the level, what is the total retarding force? What power is required to drive at 72 km h^{-1} up a 10% gradient? What is the gradient of the hill if the car coasts (i.e. uses no power) down at a constant 72 km h^{-1}?

2.9 A bullet of mass m, traveling horizontally at speed u, strikes and comes to rest in a stationary wooden block of mass M. The block is suspended as shown (Figure 2.10). Obtain an expression for the maximum height h, attained by the block after absorbing the impact of the bullet.

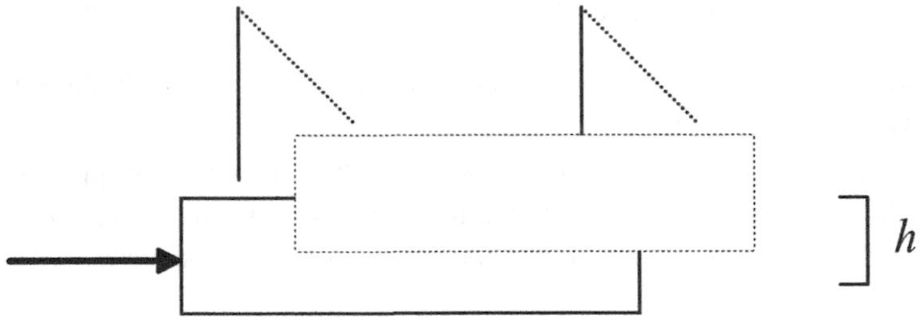

Figure 2.10 Figure for Problem 2.9.

2.10 Two astronauts, masses 40 kg and 120 kg and in need of exercise, leave their spaceship for a game of tug of war. They position themselves one at each end of a 1000 m long massless rope. Then they both pull themselves along the rope with a constant force of 60 N. What is the tension in the rope? How long does the exercise last before the astronauts meet each other? What is their relative speed when they meet? Assuming that neither astronaut is tied on to the rope, what would happen if the heavy astronaut were strong enough to pull with a force of three times the lighter one?

2.11 The balance shown in Figure 2.11 is suspended at its mid-point from the roof of a stationary lift cabin (assume the Earth to be an inertial frame and assume a constant gravitational acceleration g). The mass M_1 is chosen so that the balance arm remains horizontal. Assuming a massless, frictionless pulley, and massless, inextensible strings,

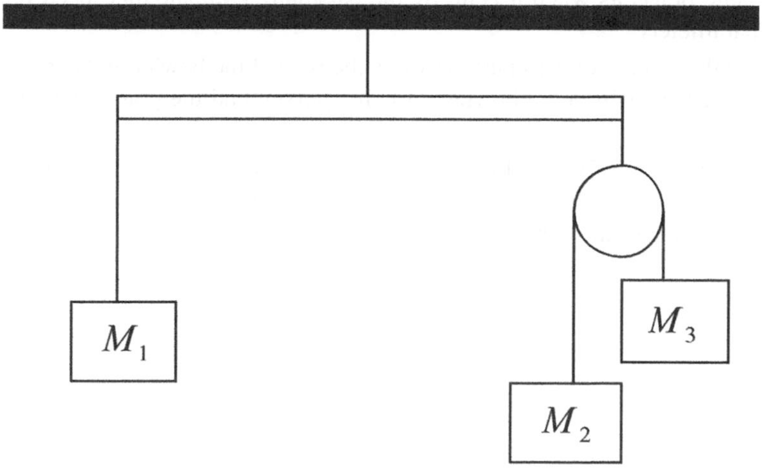

Figure 2.11 Figure for Problem 2.11.

obtain expressions for: (i) the acceleration of the mass M_2, and (ii) the tension in the string joining M_2 and M_3. Hence show that $M_1 = 4M_2M_3/(M_2 + M_3)$. The lift cabin is now accelerated upwards at a constant acceleration g. What does an observer (travelling in the lift) obtain for (i) and (ii) above? Does the balance remain horizontal?

2.12 (This one's a bit harder!) A prisoner of mass 70 kg escapes from a cell window by sliding down a rope attached at the top end to a hook which can support a pull of 600 N and no more. The rope is 15 m long and just reaches the ground. Estimate the minimum speed with which the prisoner can land on the ground if the rope (i) is massless; (ii) has a mass of 10 kg.

2.13 (a) Show that for a projectile launched with speed v and elevation α, the range, R, and the time taken to reach R, are respectively given by

$$R = \frac{v^2}{g} \sin 2\alpha \text{ and } t_R = \frac{2v}{g} \sin \alpha \, .$$

(b) In a cricket match a bowler at point A, delivers the ball to the batsman at point B. The distance AB is $w = 20$ m. The batsman then hits the ball with speed 20 m s^{-1} such that its horizontal velocity component is towards the bowler. You may neglect air resistance and any other extraneous factors. By eliminating α from the above expressions, draw a graph of R vs. t_R, marking clearly the point of maximum range.

(c) For which value of α does this maximum value correspond? Evaluate the maximum range for the given parameters.

(d) When α is given by this value, show that the bowler will just manage to catch the ball if he pursues it with speed $u = \left(v^2 - wg\right)/\sqrt{2}v$ immediately after it is hit (you may assume that when the bowler attempts to take the catch it is at the same height as when it is hit by the batsman). Evaluate this speed for the given parameters.

(e) On the same sketch as part (a), draw the path of the bowler, indicating clearly the point from which the bowler starts his pursuit and the point where he takes the catch.

(f) Comment briefly on whether or not the batsman will be caught out if he strikes the ball with α slightly less than the value for maximum range, whilst the bowler pursues the ball with the same speed as in part (d).

3

Oscillatory Motion

3.1 Introduction

As we saw in the previous chapter whenever a system is disturbed from its equilibrium state then, providing the disturbance is not too large, the restoring force is always proportional to the displacement from equilibrium about which the system then oscillates. The situation is generic to all near-equilibrium systems and is worth studying in detail in this chapter as a special case of one-dimensional motion. First, we want to solve the equation of simple oscillatory motion in a mathematically rigorous fashion for which we will introduce a little theory about the differential equations we are trying to solve. Then we want to add some realism to the model by including damping, and then the possibility of an external periodic force. The latter problem will require new mathematical techniques, so we will solve it in two ways. First, by a geometrical method (rotating arrow) and then by a more powerful technique using complex numbers. The resulting solutions will predict many new effects that are pervasive in other areas of physics, the most notable being the phenomenon of resonance, whereby the amplitude of oscillations can increase dramatically when the frequency of the external force approaches the natural frequency of the system. Finally, we consider the situation when two oscillatory systems are brought together and allowed to interact, yielding coupled oscillations.

It is worth stressing that all the situations considered in this chapter are *models*, exemplars of many oscillatory systems in physics. The analogue is exact for many electrical circuits wherein the ideas discussed find wide application. Even when the analogue is not perfect the results can still be effective – electrons in atoms are not really connected to the nuclei by springs, but a classical oscillator model provides a fair description.

3.2 Prototype Harmonic Oscillator

We begin with what is essentially the problem analysed in Section 2.10 of the previous chapter. Consider the prototype oscillator model shown in Figure 3.1. The mass, resting on a frictionless surface, is connected to a rigid support via a spring. When drawn to the right the distance from its equilibrium position is identified as x. Since it is not too far from equilibrium, the elastic limit is not exceeded and therefore the restoring force acting to the left is given by $(-kx)$, where k is the spring constant (units $\mathrm{N\,m^{-1}} = \mathrm{kg\,s^{-2}}$). Applying Newton's second law, the equation of motion is $m\ddot{x} + kx = 0$ or, dividing through by the mass and defining $\omega_0^2 \equiv k/m$,

$$\ddot{x} + \omega_0^2 x = 0 \,. \tag{3.1}$$

Classical Mechanics – Second Edition From Newton to Einstein: A Modern Introduction Martin W. McCall

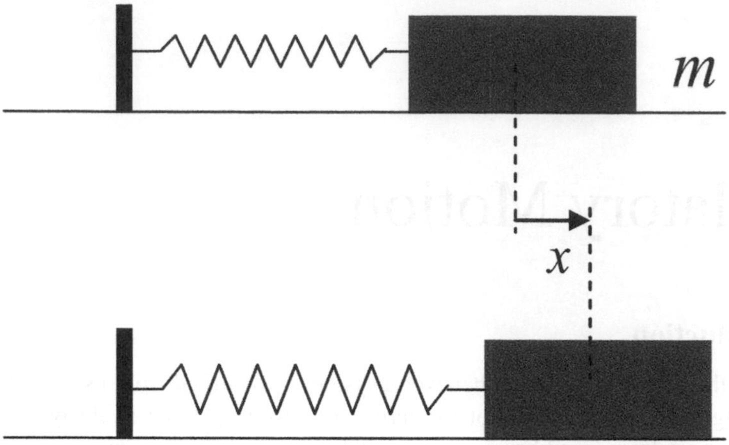

Figure 3.1 Prototype mass–spring harmonic oscillator.

Note that the parameter ω_0 has been introduced to emphasise that the motion depends on just one number, and since k/m is definitely positive we make it the square of this parameter. The physical significance of ω_0 will be clear once we have solved the equation of motion, Equation (3.1). Equation (3.1) is the equation of **Simple Harmonic Motion** (SHM), but in order to solve it rigorously, we need to know some facts about differential equations.

3.3 Differential Equations

The sorts of differential equations (DE's) we are considering are of the form

$$a\ddot{x} + b\dot{x} + cx = d(t) , \tag{3.2}$$

where there is only one dependent variable (space) and only one independent variable (time). For coupled oscillators we will relax the condition of only one dependent variable, but the results quoted below will still hold. The fact that there is only one independent variable means that we are dealing with **ordinary** (as opposed to partial) **differential equations**. We further assume that the coefficients (a, b, c) are independent of time, although d is either zero or time-dependent.[1]

First some terminology:

1. The **order** of the differential equation is its highest derivative, that is $m\ddot{x} + kx = 0$ is a second-order DE in time.
2. If each derivative occurs linearly, (that is no terms like $\dot{x}^2$ or x^3) then the DE is said to be **linear**, that is $m\ddot{x} + kx = 0$ is a **linear** DE. Note that for these purposes a term in x is regarded as a zeroth derivative. Note that the case of drag alluded to at the end of the previous chapter is an example where the equation of motion is *nonlinear*: $m\ddot{x} = -D(\dot{x})^2$.

[1] The possibility that $d \neq d(t)$ can be excluded as this simply represents a shift in the variable x, which can be eliminated via the transformation $X \equiv x - d/c$.

3. Collect all terms in $x, \dot{x}, \ddot{x}$, etc. onto the left as in Equation (3.2). If the right-hand side is independent of time (i.e. $d(t) = 0$) then the DE is termed **homogeneous**. If it is time-dependent, then it is referred to as **inhomogeneous**. For example, $m\ddot{x} + kx = 0$ is homogeneous, but $m\ddot{x} + kx = \cos(\omega t)$ is inhomogeneous.

Now some facts about the solutions of differential equations:

1. If $x_1(t)$ and $x_2(t)$ are both solutions to a linear DE, then $x_1(t) + x_2(t)$ is also a solution.
2. The general solution of a second-order DE contains two arbitrary constants.
3. The general solution of an inhomogeneous equation consists of **a particular solution** plus the general solution to the corresponding homogeneous equation.

Property 1 you can prove for yourself as an exercise. Property 2 is at least plausible since in finding the general solution to a second-order equation we will have to integrate twice with respect to time, so there will be two arbitrary constants of integration. Property 3 is for reference at this stage – we will not need it until we consider forced oscillations in Section 3.9.

With these definitions and facts we can now find the general solution to Equation (3.1) and the other differential equations considered in this chapter. Our strategy will perhaps be unfamiliar: we will propose a solution based on physical grounds, checking that it obeys the differential equation. Then we will invoke properties 1–3 above to ensure that it is the general solution.

3.4 General Solution for Simple Harmonic Motion

We saw in the previous chapter that when a system is disturbed near equilibrium it oscillates. Moreover, since the force is conservative, the energy is constant and since there is no dissipation, it oscillates with constant **amplitude**. We are thus led to propose a solution of the form

$$x_1 = A' \cos \omega t \, , \tag{3.3}$$

where A' is an arbitrary constant (the prime is for later convenience). Simple substitution of this solution into Equation (3.1) shows that this *is* a solution provided $\omega = \omega_0$. We emphasise that this is only *one* solution, by subscripting the x with a 1. It cannot be the general solution since it only contains a single arbitrary constant, A'. Similarly another, equally valid, solution is

$$x_2 = B' \sin \omega_0 t \, , \tag{3.4}$$

where B' is another arbitrary constant, and we have this time anticipated the necessary condition $\omega = \omega_0$. But now we can invoke Property 1 to find a third solution, namely

$$x = x_1 + x_2 = A' \cos \omega_0 t + B' \sin \omega_0 t \, . \tag{3.5}$$

Now this contains two arbitrary constants and, by Property 2, it *must* be the general solution.[2]

[2]You might argue that we could equally have written $x_2 = B' \cos \omega_0 t$, but adding this to x_1 merely gives a solution similar to x_1 with amplitude $A' + B'$. The solutions $x_1 = A' \cos \omega_0 t$ and $x_2 = B' \sin \omega_0 t$ are *linearly independent* – a feature we will not discuss further.

Table 3.1 Terms used in oscillatory motion.

Term	Symbol	Unit
Amplitude	A	m
Angular frequency	ω_0	rad s^{-1}
Phase	$(\omega t + \phi)$	rad
Phase constant	ϕ	rad
Period	$T_0 = 2\pi/\omega_0$	s
Frequency	$\nu_0 = 1/T_0$	s^{-1} = Hz

By simple trigonometry the solution of Equation (3.5) can be recast in the form

$$x = A\cos(\omega_0 t + \phi) \, , \tag{3.6}$$

where $\phi = \tan^{-1}(-B'/A')$ and $A = (A'^2 + B'^2)^{1/2}$ are the new pair of arbitrary constants.

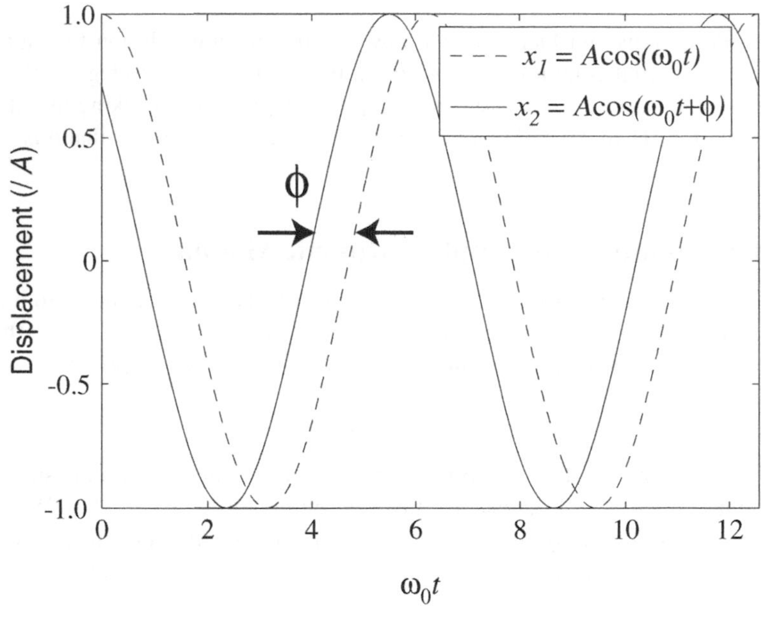

Figure 3.2 A specific solution and the general solution to the SHM problem.

The form of Equation (3.6) is the one used most frequently, and the graph of this solution is given in Figure 3.2. Note that by plotting the solution as a function of $\omega_0 t$, rather than t, the units of the ordinate are radians, and it is easy to interpret the constant ϕ: it is just the difference in phase between the general solution, Equation (3.6), with respect to the particular solution, Equation (3.3) (also plotted). The terms we have introduced so far, together with a few related quantities, are summarised in Table 3.1. Note that the period is the time required to cycle the oscillation phase through a full 2π radians. Often *angular frequency* is abbreviated to just *frequency*.

How are the arbitrary constants corresponding to the oscillation amplitude, A, and phase constant, ϕ, to be determined? Normally we are told two facts about the state of the body at time $t = 0$, namely the body's initial position and initial velocity, which together fix A and ϕ. These facts are generally referred to as **initial conditions**.

Let us apply these ideas with a simple example.

A 2 kg mass is attached to horizontal spring and displaced 1 mm from equilibrium by a force of 5 N, and then released from rest. What is the subsequent motion?

If the static extension is x_0, then $F = -kx_0 \quad \Rightarrow k = 5/0.001 = 5000$ N m^{-1}. Also, $\omega_0^2 = k/m = 5000/2 \quad \Rightarrow \quad \omega_0 = 50$ rad s^{-1}. Now applying the initial conditions to the general solution $x = A\cos(\omega_0 t + \phi)$:

$$x = x_0 \text{ at } t = 0 \quad \Rightarrow \quad x_0 = A\cos\phi\,,$$

$$\dot{x} = 0 \text{ at } t = 0 \quad \Rightarrow \quad 0 = -\omega_0 A\sin(\phi)\,.$$

Between these equations we deduce that $\phi = 0$ and $A = x_0$ so

$$x(t) = x_0\cos(\omega_0 t) = 10^{-3}\cos 50t. \tag{3.7}$$

3.5 Energy in Simple Harmonic Motion

We now examine SHM further using energy concepts. Since the restoring force is conservative, we can define a potential energy, and further, that the total mechanical energy, kinetic plus potential, is a conserved quantity. We have proved these ideas in general previously. Let us see how it all checks out for SHM.

The conservative restoring force in SHM is $F = -kx$, which can be derived from a potential function according to $-kx = -dU/dx$. Hence

$$U = -\int F dx = \frac{1}{2}kx^2. \tag{3.8}$$

The total mechanical energy is therefore

$$E = K + U = \frac{1}{2}mv^2 + \frac{1}{2}kx^2. \tag{3.9}$$

The time-independence of E can be checked by substituting the general solution $x(t)$ from Equation (3.6):

$$E = \frac{1}{2}m\left[-\omega_0 A\sin(\omega_0 t + \phi)\right]^2 + \frac{1}{2}k\left[A\cos(\omega_0 t + \phi)\right]^2$$

$$= \frac{1}{2}m\omega_0^2 A^2\sin^2(\omega_0 t + \phi) + \frac{1}{2}kA^2\cos^2(\omega_0 t + \phi).$$

But $\omega_0^2 = k/m$, so

$$E = \frac{1}{2}kA^2 = \frac{1}{2}m\omega_0^2 A^2\,, \tag{3.10}$$

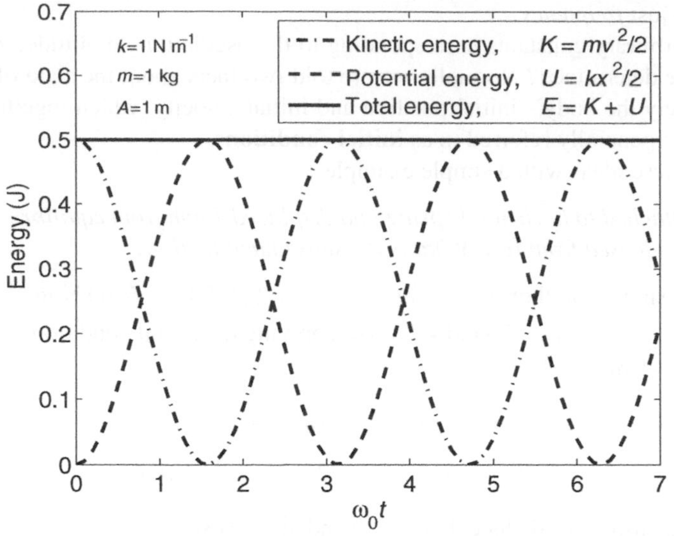

Figure 3.3 Time dependence of the kinetic and potential energy in SHM when $\phi = 0$.

which is constant as claimed. The kinetic energy and the potential energy are time-dependent as shown in Figure 3.3 (for $\phi = 0$), but their sum representing the total energy is constant. The potential energy and the kinetic energy are exchanged four times every oscillation period, whilst the total mechanical energy ($= m\omega_0^2 A^2/2$) corresponds to the potential energy at maximum displacement $x = A$, or the kinetic energy at the equilibrium position $x = 0$.

The model developed in this section predicts that the oscillations will continue forever, and whilst an oscillator may go through many cycles without significant energy loss, eventually friction will cause the amplitude to decay and for the oscillations to die out. We consider this more realistic model next.

3.6 Damped Oscillations

We now refine our model system to that shown in Figure 3.4 in which the mass oscillates within a viscous medium, causing the vibrations to decay, or to be **damped**. We considered resistive motion without oscillations in Section 2.13. Similarly here we make the simplest possible assumption concerning the nature of the viscous force, namely that it is proportional to the speed of the mass and opposes its motion. As in Equation (2.39) we write

$$F_{\text{resistance}} = -bv = -b\dot{x}. \tag{3.11}$$

Including this extra term in Newton's second law gives

$$m\ddot{x} = -b\dot{x} - kx, \tag{3.12}$$

or, dividing through by the mass,

$$\ddot{x} + \gamma\dot{x} + \omega_0^2 x = 0, \tag{3.13}$$

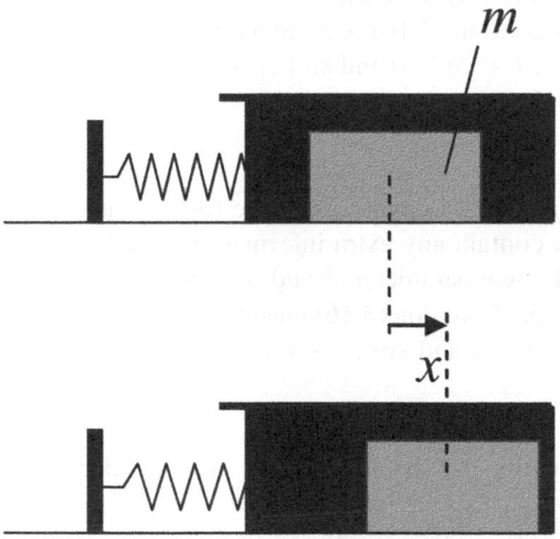

Figure 3.4 Model system for damped oscillations: a spring connected to a mass surrounded by a viscous medium.

where $\gamma \equiv b/m$. How are we going to solve Equation (3.13)? If this were a maths book we would just develop the necessary mathematical machinery and solve it. On the other hand, a simple-minded physicist, who missed the course on differential equations, might be tempted to reiterate the experimental observation that the mass oscillates and the oscillations decay with time, without actually attempting a solution. We are going to take a middle course combining intuition and rigour. Knowing that the motion consists of damped oscillations prompts us to propose the solution of Equation (3.13) as

$$x = Ae^{-qt}\cos(\omega t + \phi), \tag{3.14}$$

the amplitude now set to decay exponentially. That the mass oscillates is expressed through the factor $\cos(\omega t + \phi)$, just as in Equation (3.6), but in this case the oscillation frequency has still to be determined and is *not* given by $\omega = \omega_0 = \sqrt{k/m}$. Regarding the exponential decay we reason as follows: if the spring were absent, then after setting $v = \dot{x}$, the equation of motion would read

$$\dot{v} + \gamma v = 0. \tag{3.15}$$

The solution of this is $v = v_0 \exp(-\gamma t)$, where v_0 is the initial velocity. Integrating again with respect to t, yields $x = x_0 + (v_0/\gamma)\,[1 - \exp(-\gamma t)]$, where x_0 is the initial position. The effect of damping is therefore to exponentially inhibit the velocity and the position of the body. In our proposed solution we anticipate this feature by writing the overall amplitude in Equation (3.14) as a decaying exponential, with **decay constant** q, to be determined.

Substituting Equation (3.14) into Equation (3.13) and grouping the sine and cosine terms together results in

$$\left(-\omega^2 + q^2 - \gamma q + \omega_0^2\right) Ae^{-qt}\cos(\omega t + \varphi) + (2q\omega - \gamma\omega)\, Ae^{-qt}\sin(\omega t + \varphi) \equiv 0. \tag{3.16}$$

Why have we grouped the terms in this way, and why have we used '≡' rather than '=' in Equation (3.16)? The point is that Equation (3.16) is an identity[3] *true for all time*, t. If we first choose a time which makes $\cos(\omega t + \phi) = 0$ and $\sin(\omega t + \phi) = 1$ (e.g. $t = (\frac{\pi}{2} - \phi)/\omega$ will do the job) then Equation (3.16) becomes

$$(2q\omega - \gamma\omega) = 0 \tag{3.17}$$

and consequently either $\omega = 0$ or $q = \gamma/2$. Leaving aside the former case for the moment (in fact we will find that it does not contain any extra information), the latter case fixes the damping constant q in terms of the known parameters b and m, since $\gamma = b/m$.

Half way there, and remembering that Equation (3.16) must be true for all t, we can now set $t = -\phi/\omega$, which makes $\cos(\omega t + \phi) = 1$ and $\sin(\omega t + \phi) = 0$. This time Equation (3.16) becomes

$$\left(-\omega^2 + q^2 - \gamma q + \omega_0^2\right) = 0. \tag{3.18}$$

Using the fact that $q = \gamma/2$, Equation (3.18) determines that the damped oscillation frequency is given by

$$\omega^2 = \omega_0^2 - \gamma^2/4. \tag{3.19}$$

The damped oscillation frequency is therefore less than the natural frequency ω_0, or a damped system oscillates with a longer period than the equivalent undamped system. The negative square root of Equation (3.19) yields a negative frequency, which can be interpreted in the solution as equivalent to replacing t with $-t$. Such a time reversal does not yield any new information and the negative frequency is usually ignored.

Now that we have found out q and ω, the solution may be written as

$$x = Ae^{-\gamma t/2}\cos(\omega t + \phi) \tag{3.20}$$

with ω given by Equation (3.19). Just as in the undamped case, the constants A and ϕ are determined through the initial conditions, usually stated as the position and velocity of the mass at $t = 0$. Since these constants are different for each specific case under consideration, they are precisely the arbitrary constants associated with the general solution of the second-order DE, Equation (3.13). We can therefore assert definitively that Equation (3.20) is the general solution. An example of damped oscillations is given in Figure 3.5. For the chosen parameters ($\omega_0 = 1\mathrm{rad\ s}^{-1}$, $\gamma = 0.1\ \mathrm{s}^{-1}$) the damped oscillation frequency is $\omega = (1^2 + 0.1^2/4)^{1/2} = 1.0012\ \mathrm{rad\ s}^{-1}$, or in other words very close to ω_0. This is reflected in the figure by the fact that both the damped and the undamped oscillations appear to be in phase. Is it ever possible to perceive the change in frequency in the damped case? For the phase of the damped and undamped oscillations to be out of phase by $\pi/2$, we would have to wait a time, t, given by

$$(\omega_0 - \omega)\,t = \frac{\pi}{2}.$$

When $\omega_0 \gg \gamma$ as here, a first-order expansion of Equation (3.19) yields $\omega \approx \omega_0(1 - \gamma^2/8\omega_0^2)$, so that the time to achieve phase slippage by a quarter period is, for the parameters given

$$t = \frac{4\pi\omega_0}{\gamma^2} = 1258\ \mathrm{s}.$$

[3] You can appreciate the difference between an identity and an equation as follows. The solution of the *equation* $ax^2 + bx + c = 0$ is $x = (-b \pm \sqrt{b^2 - 4ac})/2a$. The solution of the *identity* $ax^2 + bx + c \equiv 0$ is $a = b = c = 0$, as only in this way can the expression be true for all x.

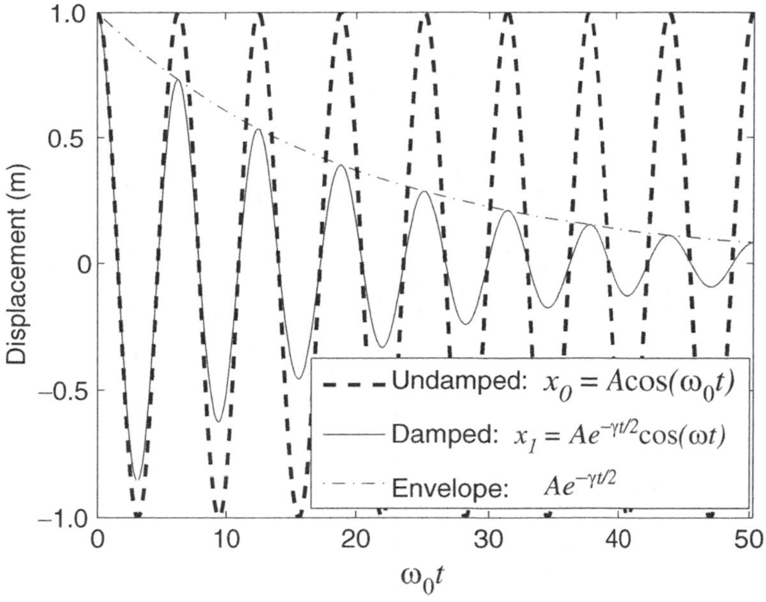

Figure 3.5 Plots of damped (Equation (3.20)) and undamped (Equation (3.6)) oscillations. Parameter values are given in the text.

A plot of the damped and undamped oscillations starting at this time is shown in Figure 3.6 and indeed the damped oscillations are ahead of the undamped ones by a quarter of a period. But look what's happened to the amplitude meanwhile. It has reduced to $\exp(-0.1 \times 1258/2) = 5.1 \times 10^{-28}$ times its initial value! Indeed, the height of the undamped oscillations has had to be scaled down to be comparable to the damped oscillations in the figure. The amplitude of the damped oscillation is reduced by a factor of $\exp(-2\pi\omega_0/\gamma)$ once a significant phase difference has been established. This factor is evidently $\ll 1$. The message is that the phase difference is very hard to detect in practice, and for most practical purposes $\omega \approx \omega_0$.

3.7 Light Damping – the *Q* Factor

There are two natural timescales associated with damped SHM:

1. $T_0 = 2\pi/\omega_0 = 2\pi\sqrt{m/k}$ is the **natural period**, or the period with which the system would oscillate in the absence of damping.
2. $T_D = 1/\gamma = m/b$ is the **decay time constant**. We have seen that the amplitude decays to $1/e$ of its original value after a time $2T_D$.

The ratio of these two time periods gives a dimensionless number, which when multiplied by 2π, is called the **Q value** of the oscillator:

$$Q = \frac{\omega_0}{\gamma} = \frac{2\pi T_D}{T_0} = \sqrt{\frac{km}{b^2}}\,. \tag{3.21}$$

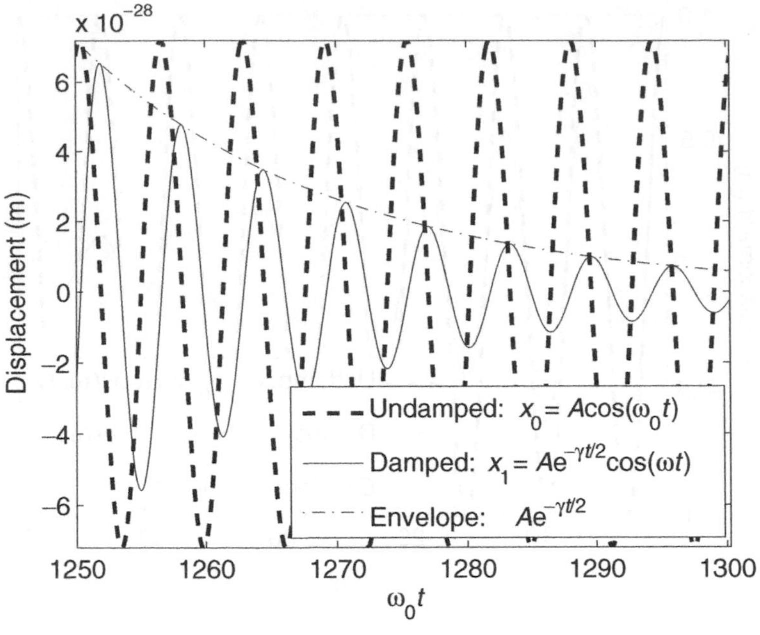

Figure 3.6 Plots of damped (Equation (3.20)) and undamped oscillations (Equation (3.6)) from $t = 1250$ s to 1290 s. Note that the curve for undamped oscillations is scaled down by a factor of $\exp(625t)$. Parameter values are given in the text and as for Figure 3.5.

Table 3.2 Properties of typical oscillatory systems.

	Q	ω_0 (rad s^{-1})	γ (s^{-1})
Mechanical pendulum	10^2	1	10^{-2}
Tuning fork	10^4	10^3	10^{-1}
FM tuner	10^4	10^8	10^4
Excited atom	10^7	10^{15}	10^8

The Q value gives a measure of the strength of the damping of the oscillating system, a 'high Q' oscillator (that is $\omega_0 \gg \gamma$) going through many cycles before the oscillation amplitude has significantly decayed. The Q of some common oscillating systems, together with their corresponding natural frequencies and damping rates are shown in Table 3.2, which also serves to highlight the breadth of applicability of oscillator theory.

Physicists actually use different definitions of Q, but fortunately they all converge when Q is sufficiently high. Since this is the regime in which the system Q is usually discussed we can safely ignore these distinctions and stick to the definition given in Equation (3.21).

Apart from quantifying the lightness of the damping, does the Q factor have any other physical significance? Since the energy stored in an oscillating system is proportional to the square of the amplitude (see Equation (3.10))

$$E \propto A^2 \exp(-\gamma t)$$

or, in terms of $E(0)$, the initial system energy

$$E(t) = E(0)\exp(-\gamma t). \qquad (3.22)$$

After one further oscillation period $T = 2\pi/\omega$, the stored energy becomes

$$E(t+T) = E(0)\exp\left[-\gamma(t+T)\right] = E(t)\exp\left(-\gamma T\right). \qquad (3.23)$$

With light damping $T \approx T_0$ and $\gamma T \ll 1$, so that the energy lost by the system between time t and $t+T$ becomes

$$\Delta E = E(t)\left[1 - \exp(-\gamma T)\right] \approx E(t)\cdot\gamma\cdot T_0 = \frac{2\pi}{Q}\cdot E. \qquad (3.24)$$

In other words the *fraction* of energy lost per cycle $\Delta E/E$, is just $2\pi/Q$. Since each cycle consists of 2π radians this motivates an equivalent useful definition of Q:

- The Q, or **quality factor**, of a damped oscillating system is the energy stored by the system divided by the energy lost per radian of oscillation, $Q = \frac{1}{2\pi}\left(\frac{E}{\Delta E}\right)$.

3.8 Heavy Damping and Critical Damping

So far we have presumed that the damping has been light so that $\omega_0 > \gamma/2$ results in a real oscillation frequency ω. There is, however, no physical restriction on the magnitudes of ω_0 and γ, since they are determined by the spring constant k, the resistive constant b and the mass m. With some trepidation, therefore, we must confront the possibility of an imaginary oscillation frequency resulting when $\gamma/2 > \omega_0$. What can an imaginary frequency mean? After all, no matter what the mathematics throws up, physically the displacement observed in an oscillating system must be real. Sometimes it's best just to let the maths lead the way and carry on regardless. First we note that the phase constant, ϕ, appearing in Equation (3.20), merely charts the phase of the oscillation, and can be made to vanish by judicious choice of time origin. We will imagine that this has been done so that $\phi = 0$. The frequency meanwhile is given by

$$\omega = \sqrt{\omega_0^2 - \left(\frac{\gamma}{2}\right)^2} = i\sqrt{\left(\frac{\gamma}{2}\right)^2 - \omega_0^2} = i\alpha \qquad (3.25)$$

where α is real. Now we can look at how the solution (Equation (3.20)) behaves by noting that

$$\cos(i\alpha t) = \frac{1}{2}\left[e^{i(i\alpha t)} + e^{-i(i\alpha t)}\right] = \cosh(\alpha t) \qquad (3.26)$$

so that now

$$x(t) = Ae^{-\frac{\gamma}{2}t}\left(e^{-\alpha t} + e^{\alpha t}\right)/2 = \frac{A}{2}\left(e^{-\alpha_+ t} + e^{-\alpha_- t}\right), \qquad (3.27)$$

where $\alpha_\pm \equiv \pm\alpha + \gamma/2 = \pm(\gamma^2/4 - \omega_0^2)^{1/2} + \gamma/2$. The displacement is real after all. However, the system no longer oscillates but decays to equilibrium exponentially with two rate constants, α_+ and α_-, and is referred to as being **heavily damped**, or **overdamped**. The decay constants are both positive for all possible parameter values, which is fortunate as otherwise at least part of the solution Equation (3.27) would *increase* exponentially, and we cannot have that. The reciprocals of α_+ and α_- define two corresponding time constants,

$$T_{\pm} = \left[\frac{\gamma}{2} \pm \sqrt{\left(\frac{\gamma}{2}\right)^2 - \omega_0^2}\right]^{-1}. \tag{3.28}$$

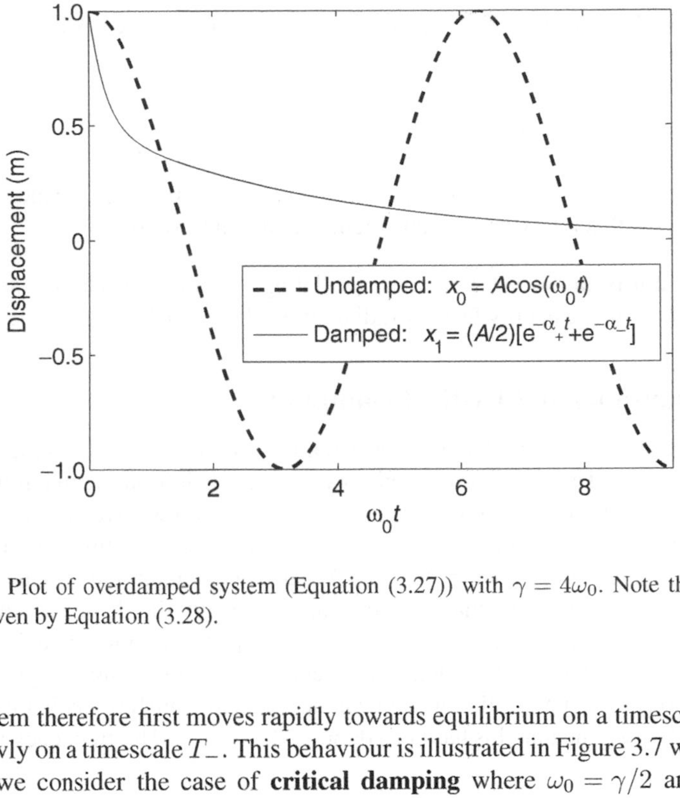

Figure 3.7 Plot of overdamped system (Equation (3.27)) with $\gamma = 4\omega_0$. Note the two decay time constants given by Equation (3.28).

The system therefore first moves rapidly towards equilibrium on a timescale T_+, and then decays slowly on a timescale T_-. This behaviour is illustrated in Figure 3.7 wherein $\gamma = 4\omega_0$.

Finally we consider the case of **critical damping** where $\omega_0 = \gamma/2$ and $\omega = 0$.[4] Now $T_+ = T_- = 2/\gamma$ and the solution is simply

$$x(t) = Ae^{-\frac{\gamma}{2}t} \tag{3.29}$$

the system decaying towards equilibrium more rapidly than in the overdamped case, but still just fails to oscillate as shown in Figure 3.8. Car suspension systems are often critically damped to ensure the most comfortable ride over bumpy road surfaces.

[4]The case of $\omega = 0$ considered here is a special case of zero frequency given as a possible solution to Equation (3.17). There, had we pursued $\omega = 0$ as a possible solution, then q determined by Equation (3.18) would have turned out to be complex, containing both decaying and oscillating parts. The imaginary component corresponds to what we have calculated with $\omega \neq 0$, so we have not lost any generality in deciding to ignore $\omega = 0$ as a possible solution to Equation (3.17).

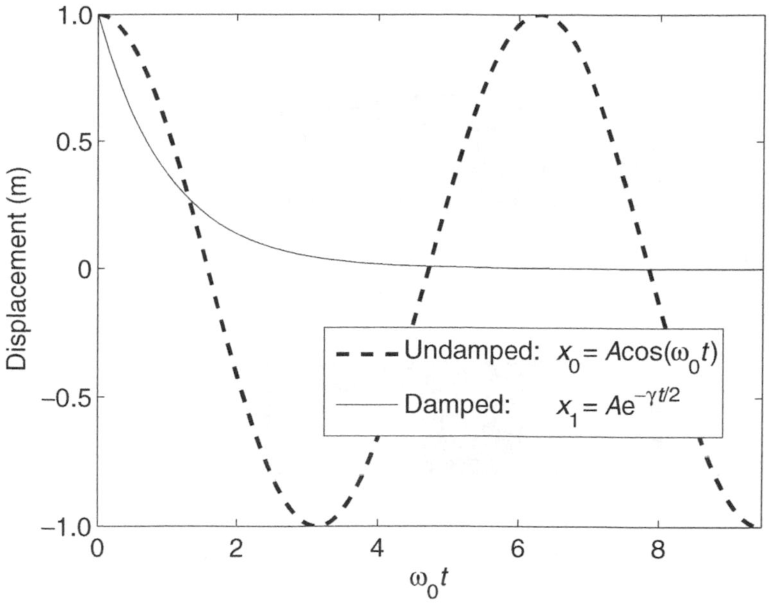

Figure 3.8 Return to equilibrium of a critically damped system where $\gamma = 2\omega_0$.

3.9 Forced Oscillations

A pendulum clock and a radio set are simple examples where we want to maintain an oscillation in the presence of damping. This can only be achieved by applying some external force and our dash-pot model oscillator system receives its final refinement in Figure 3.9 wherein the support is presumed to be subjected to a periodic force given by

$$F_{\text{ext}} = F_0 \cos(\omega_d t). \tag{3.30}$$

In Equation (3.30) both F_0, the force amplitude, and ω_d, the driving angular frequency of the applied force, are considered fixed. Three forces now act on the mass: (i) the return force of the spring, (ii) the resistive force and (iii) the externally applied periodic force, so that Newton's second law reads

$$m\ddot{x} = -kx - b\dot{x} + F_0 \cos(\omega_d t)\,. \tag{3.31}$$

After dividing through by the mass and introducing ω_0 and γ as previously, we obtain the equation of motion

$$\ddot{x} + \gamma\dot{x} + \omega_0^2 x = \frac{F\cos(\omega_d t)}{m}. \tag{3.32}$$

Equation (3.32) now represents a significant mathematical challenge. However, we will continue in the spirit of the heuristic approach through which we successfully solved the unforced oscillator systems, namely to propose a solution on physical grounds, and to then check at the end that it satisfies all the conditions for the general solution. In this regard we

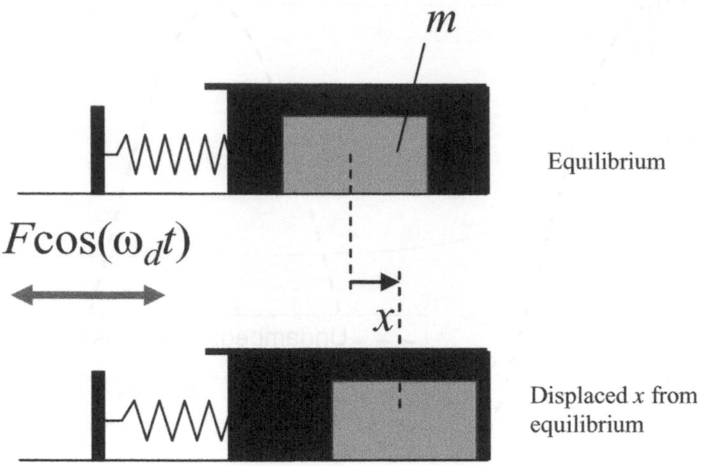

Figure 3.9 Forced damped harmonic oscillator model.

note that since the right-hand side of Equation (3.32) is now time-dependent, the equation is inhomogeneous and Property 3. in Section 3.3 will therefore be relevant.

What kind of solution, $x(t)$, do we expect? If we imagine pumping away at the support, we expect the mass to oscillate at the same frequency ω_d and we are prompted to write

$$x(t) = A\cos(\omega_d t + \phi) \tag{3.33}$$

as a potential solution to Equation (3.32). Notice that we take account of the fact that the mass cannot respond *instantaneously* by including the phase constant ϕ in Equation (3.33). A very useful representation of Equation (3.33) is to regard $x(t)$ as the x-component of an arrow of length A, rotating anticlockwise in the x–y plane as shown in Figure 3.10. In this representation $(\omega_d t + \phi)$ is the instantaneous angle that the arrow makes with the x-axis, and ϕ is the initial value of this angle. By differentiating Equation (3.33) we find similar rotating arrow representations for all the terms in Equation (3.32):

$$\dot{x} = -\omega_d A\sin(\omega_d t + \phi) = \omega_d A\cos(\omega_d t + \phi + \frac{\pi}{2}), \tag{3.34}$$

and

$$\ddot{x} = -\omega_d^2 A\cos(\omega t + \phi) = \omega_d^2 A\cos(\omega_d t + \phi + \pi). \tag{3.35}$$

We have recast both the velocity and the acceleration in a form similar to Equation (3.33), showing that (a) the velocity *leads* the displacement in phase by $\pi/2$, and (b) the acceleration is in *anti-phase* with respect to the displacement. The three terms on the left of Equation (3.32) may therefore be represented by the triad of rotating arrows shown in Figure 3.11, the lengths of which are respectively $\omega_0^2 A$, $\gamma\omega_d A$ and $\omega_d^2 A$. How does the periodic force of Equation (3.30) fit into this picture? Since it has no phase constant (and in fact *defines* the phase origin) it must be represented by an arrow pointing along the x-axis.

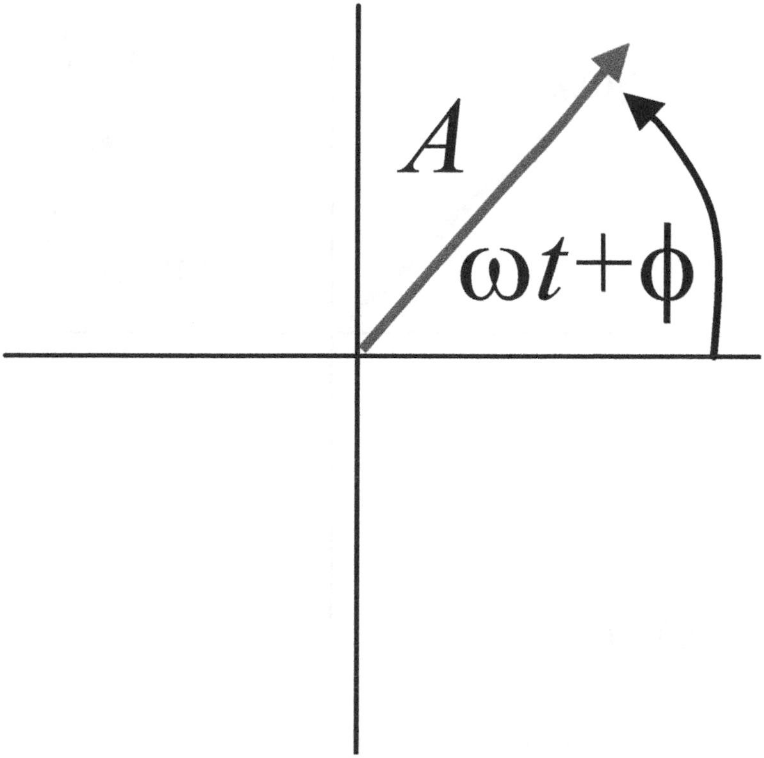

Figure 3.10 Rotating line representation of $x(t)$ in Equation (3.33).

The point of all this is that if Equation (3.32) is to be satisfied then the arrows, now considered as *vectors* in the x–y plane, *must form a closed quadrilateral.* In this way the x-components of all the vectors will add up so as to satisfy Equation (3.32). Why must the quadrilateral close? Because at some later time, as the vectors rotate what are initially the y-components of the vectors will lie along the x-axis, and now they too must satisfy Equation (3.32). The only way to satisfy this requirement is for the arrows to join up in a closed quadrilateral. A little thought shows that to maintain the required phase relationships between the arrows corresponding x, $\dot{x}$ and $\ddot{x}$ and to keep the arrow corresponding to F_{ext} lying along the x-axis is to have the arrows close up as shown in Figure 3.12 (which shows the arrows at $t = 0$). The angle ϕ must be $> \pi$, or else the vectors representing x, $\dot{x}$ and $\ddot{x}$ will never 'catch up' to close off the quadrilateral with F_{ext}.

Now as noted above the horizontal and vertical components of the quadrilateral must separately equate as in Equation (3.32) so that taking each component in turn

$$F/m = \omega_0^2 A \cos\phi - \gamma\omega_d A \sin\phi - \omega_d^2 A \cos\phi\,, \tag{3.36}$$

and

$$0 = \omega_0^2 A \sin\phi + \gamma\omega_d A \cos\phi - \omega_d^2 A \sin\phi. \tag{3.37}$$

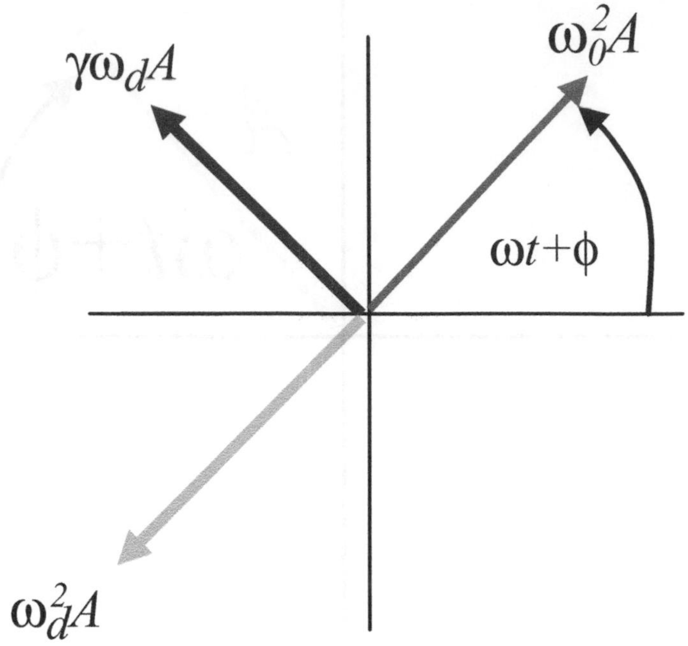

Figure 3.11 Rotating line representations the displacement, velocity and acceleration terms on the left of Equation (3.32).

These two equations may be solved simultaneously for A and ϕ yielding

$$A = \frac{F}{m} \Big/ \sqrt{(\omega_0^2 - \omega_d^2)^2 + \gamma^2\omega_d^2} \tag{3.38}$$

and

$$\tan\phi = \frac{-\gamma\omega_d}{\omega_0^2 - \omega_d^2} . \tag{3.39}$$

These are our main results. The amplitude and phase of the displacement, as calculated in Equations (3.38) and (3.39), are *determined* functions of the applied force and the model parameters. This is in marked contrast to the previous case of unforced oscillations where in Equations (3.6) and (3.20) A and ϕ were the solution's *arbitrary* constants, determined by the initial conditions. Figures 3.13 and 3.14 show plots of the oscillation amplitude and phase as functions of the driving frequency for various values of the resistive damping constant. The most striking feature of the graphs for the oscillation amplitude is the presence of a **resonance** at a frequency a little below the natural frequency, ω_0, at which point the amplitude increases markedly, the more the lighter the damping. We can calculate the resonance frequency exactly by differentiating Equation (3.38). In fact, we can save ourselves a bit of algebra by observing that A is a maximum when the denominator on the right is a minimum, and indeed we need

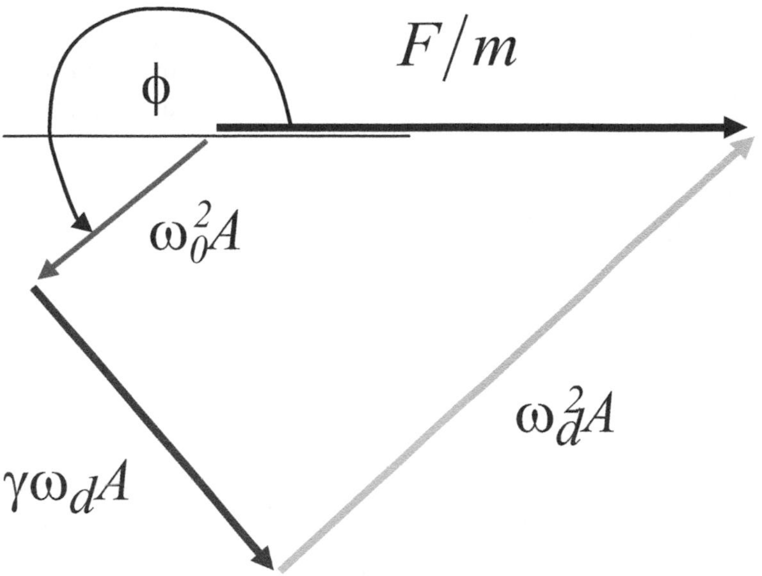

Figure 3.12 Closed quadrilateral of rotating arrows representing Equation (3.32) at $t = 0$.

only minimise the function under the square root, and set the resulting function to zero:

$$2\left(\omega_0^2 - \omega_d^2\right)(-2\omega_d) + 2\gamma^2\omega_d = 0$$

or

$$\omega_d(\text{resonance}) = \left(\omega_0^2 - \frac{\gamma^2}{2}\right)^{1/2}. \tag{3.40}$$

The graphs of Figure 3.13 show that the resonance feature is more pronounced for light damping, that is when $\omega_d \approx \omega_0$, the natural frequency. At resonance the amplitude is

$$A_{\text{max}} = \frac{F/m}{\gamma\left(\omega_0^2 - \frac{\gamma^2}{4}\right)^{1/2}} \approx \frac{F}{m\gamma\omega_0}, \tag{3.41}$$

the approximation applying when $\omega_0 \gg \gamma$. A useful form of this expression results if we note that when the drive frequency is zero ($\omega_d = 0$) the amplitude is $A_{\text{static}} = F/(m\omega_0^2)$ so that

$$A_{\text{max}} = \frac{F}{m\omega_0^2}\frac{\omega_0}{\gamma} = A_{\text{static}} \times Q \tag{3.42}$$

and the Q factor appears again, this time as the enhancement of the displacement at resonance. Indeed, the width of the resonance as measured, say, by the width of the amplitude resonance at half the maximum value, is found after some algebra to be proportional to $1/Q$, so that for a high Q oscillator system the resonance becomes narrower and sharper. As the driving frequency increases beyond resonance the amplitude becomes smaller, tending to

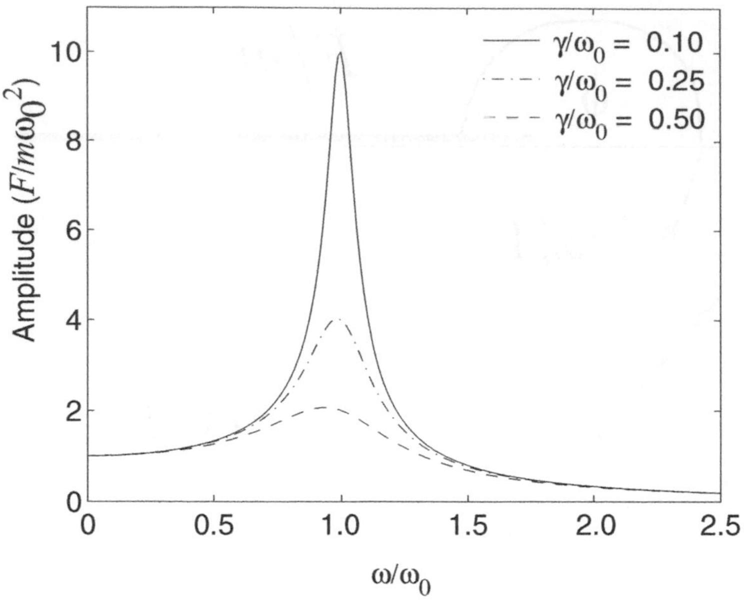

Figure 3.13 Plot of forced oscillation amplitude, A, for various values of resistive damping, γ.

zero for large ω_d. This is due to the fact that at large frequencies the mass is too heavy to respond to the rapid driving oscillations.

Sketching the phase curves requires a little thought with respect to Equation (3.39) and Figure 3.14. As the driving frequency increases away from zero, the mass, initially able to follow the driving force more or less instantaneously, starts to lag behind, and ϕ becomes negative away from zero. By how much can the displacement lag behind as ω_d increases further? It cannot slip behind by as much as 2π radians, as then effectively the displacement is in phase again with respect to the driving force. The extreme of large ω_d is therefore characterised by the displacement being in *anti-phase* with respect to the force. The transition between in-phase and anti-phase behaviour occurs when $\omega_d = \omega_0$, the transition being sharper for lighter damping.

Does Equation (3.33), with amplitude and phase given by Equations (3.38) and (3.39), constitute the general solution to the forced oscillator problem (Equation (3.32))? No! For one thing, what happened to the two arbitrary constants that we said in Section 3.3, Property 2 are necessary accompaniments to the general solution of a second-order ordinary differential equation? The amplitude and phase are no longer arbitrary as they are determined by Equations (3.38) and (3.39). But we can invoke Properties 1 and 3 to guide us towards the general solution. Property 3 says that the general solution consists of the sum of the solution to the corresponding homogeneous problem obtained when $F_{\text{ext}} = 0$, and the particular solution found when $F_{\text{ext}} = F_0 \cos(\omega_d t)$. Therefore what we have calculated so far in this section has pertained to the particular solution. Fortunately no more calculation is necessary as the solution to the corresponding homogeneous problem is precisely what we calculated in Section 3.6, Equation (3.20), for unforced damped oscillations. The general solution of the

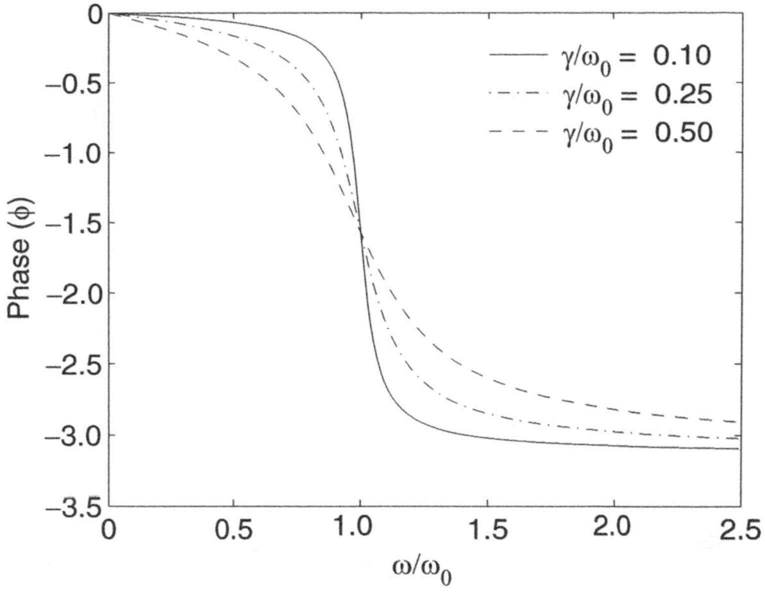

Figure 3.14 Plot of forced oscillation phase, ϕ, for various values of resistive damping, γ.

forced oscillator equation (Equation (3.32)) is therefore

$$x(t) = \underbrace{A_0 e^{-\gamma t/2}\cos(\omega t + \phi_0)}_{\text{Transient}} + \underbrace{A\cos(\omega_d t + \phi)}_{\text{Steady-state}} , \tag{3.43}$$

where the only modification we have made is to change the amplitude and phase appearing in Equation (3.20) to A_0 and ϕ_0 to distinguish them from the A and ϕ calculated in Equations (3.38) and (3.39). Now A_0 and ϕ_0 *are* the arbitrary constants associated with the initial conditions signalling that we have found the general solution. Actually the first part of the solution in Equation (3.43) decays away with time, so that if we wait long enough it is irrelevant. For this reason it is referred to as the **transient** part of the solution, the remaining term corresponding to the long term, or **steady-state** behaviour. Figure 3.15 shows the evolution of the system as per Equation (3.43) for different initial conditions. As can be seen the initial behaviour during the transient phase can be quite violent. However, the system always settles down to the same steady-state behaviour, independent of the initial conditions.

3.10 Complex Number Method

The previous section gave all the details leading up to the general solution of the forced harmonic oscillator problem given in Equation (3.43). However, we had to invoke a rather unusual solution method involving rotating arrows etc., and so in this section we are briefly going to give an elegant alternative method based on complex numbers.

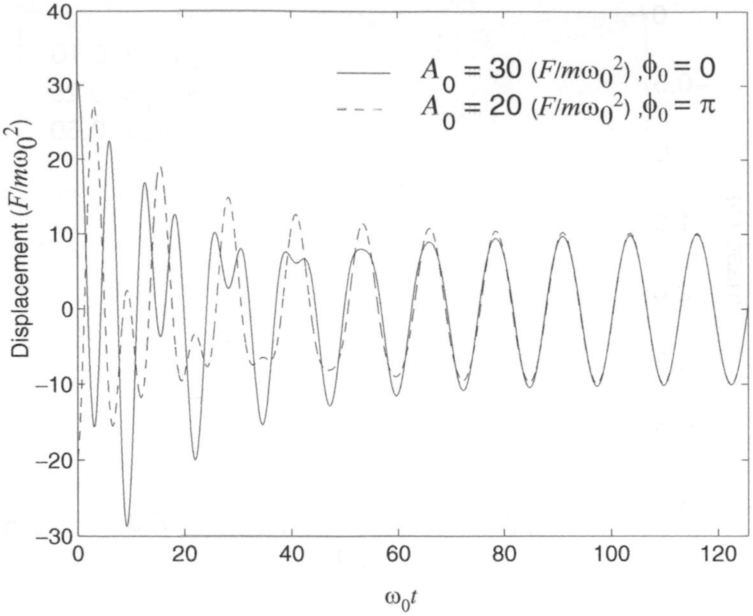

Figure 3.15 Plot of the general solution Equation (3.43) of the forced oscillator equation for different initial conditions. Parameters: $\omega_d = 0.5\omega_0$, $\gamma = 0.1\omega_0$.

The key observation is to associate the plane supporting the rotating arrow representation of $x(t)$ in Figure 3.11, with the Argand diagram supporting a complex number $\tilde{x}(t)$, defined through

$$\tilde{x} = A\exp[i(\omega t + \phi)] = \tilde{A}\exp(i\omega t) \tag{3.44}$$

where $\tilde{A}$ is the complex number $A\exp(i\phi)$ containing both the amplitude and phase of the oscillation. Figure 3.16 shows the Argand representation of $\tilde{x}(t)$, whence the actual physical displacement is evidently given by its real part:

$$x = \text{Re}(\tilde{x}). \tag{3.45}$$

We now define an associated forced oscillator equation using the complex displacement as

$$\ddot{\tilde{x}} + \gamma\dot{\tilde{x}} + \omega_0^2\tilde{x} = \frac{\tilde{F}}{m}, \tag{3.46}$$

where

$$\tilde{F} = F\exp(i\omega_d t), \tag{3.47}$$

defines a complex force. Evidently the original real forced oscillator equation, Equation (3.31), is recovered by taking the real part of Equation (3.46). Let us emphasise that this is simply a mathematical device. If we can somehow solve Equation (3.46) for $\tilde{x}(t)$ then we will be able to recover the actual displacement via Equation (3.45). The point is it turns out

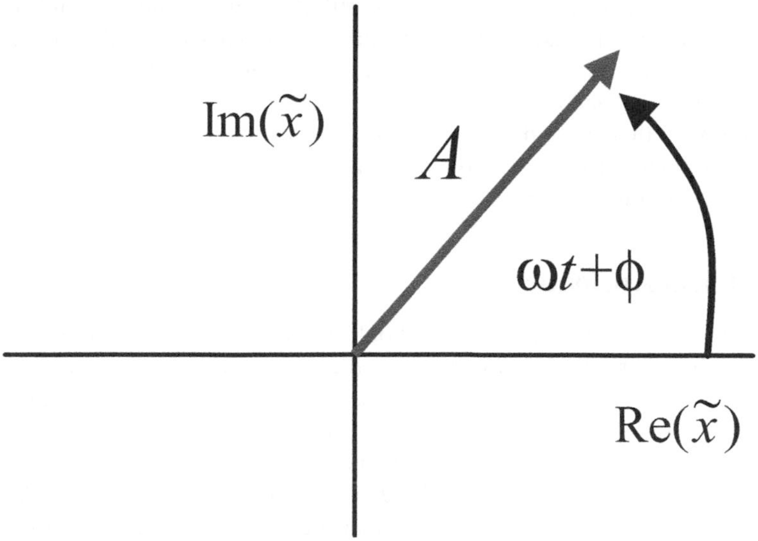

Figure 3.16 Argand representation of the complex displacement $\tilde{x}(t)$.

to be much easier to solve the associated complex oscillator equation than the original real equation! If we try the solution

$$\tilde{x} = \tilde{A}\exp(i\omega_d t)\,, \tag{3.48}$$

then after calculating the derivatives $\dot{\tilde{x}} = i\omega_d\tilde{x}$ and $\ddot{\tilde{x}} = -\omega_d^2\tilde{x}$, Equation (3.46) becomes

$$-\omega_d^2\tilde{A} + \gamma i\omega_d\tilde{A} + \omega_0^2\tilde{A} = \frac{F}{m}. \tag{3.49}$$

Immediately the complex amplitude $\tilde{A}$ is calculated as

$$\tilde{A} = A\exp(i\phi) = \frac{F/m}{[\omega_0^2 - \omega_d^2 + i\gamma\omega_d]}\,. \tag{3.50}$$

In order to dissect this into A and ϕ it is best to multiply the numerator and the denominator by the complex conjugate of the denominator resulting in

$$A\exp(i\phi) = \frac{(F/m)\left(\omega_0^2 - \omega_d^2 - i\gamma\omega_d\right)}{\left[\left(\omega_0^2 - \omega_d^2\right)^2 + \left(\gamma\omega_d\right)^2\right]}\,. \tag{3.51}$$

Equation (3.51) is a complex number of the form $a + ib$. The amplitude and phase are respectively given by $\sqrt{(a^2 + b^2)}$ and $\tan^{-1}(b/a)$. A and ϕ are therefore given by

$$A = \frac{F/m}{[(\omega_0^2 - \omega_d^2)^2 + \gamma^2\omega_d^2]^{1/2}}\,, \tag{3.52}$$

and

$$\tan\phi = \frac{-\gamma\omega_d}{\omega_0^2 - \omega_d^2} \tag{3.53}$$

exactly as before in Equations (3.38) and (3.39)! By using complex numbers we have obtained the amplitude and phase of the steady-state solution to the forced oscillator problem. As an exercise you can show that the same technique can be applied to the homogenous equation Equation (3.13) and that the oscillation frequency and the damping constant of the transient solution are then those determined from Equations (3.17) and (3.19).[5]

The complex number method can always be used to convert a linear ordinary differential equation into an algebraic equation.

3.11 Electrical Analogue

This is not a book on electronics, but we should not pass up the opportunity to use what we have learnt about forced oscillators to solve the model LCR electronic circuit. Consider the circuit in Figure 3.17. consisting of an inductance, L, a resistance, R, and a capacitor, C, all in series with an alternating voltage source $V\cos(\omega_d t)$. In terms of the charge, $q(t)$, and the current, $I = \dot{q}$, passing through each element, the voltages dropped over the inductor, resistor and the capacitor are respectively $L(dI/dt) = L\ddot{q}$, $RI = R\dot{q}$ and q/C. According to Kirchhoff's law, the sum of these voltage drops must equal the applied alternating voltage so that

$$L\ddot{q} + R\dot{q} + \frac{q}{C} = V\cos(\omega_d t). \tag{3.54}$$

Identifying $F \leftrightarrow V$, $m \leftrightarrow L$, $b \leftrightarrow R$, $k \leftrightarrow 1/C$ and $x \leftrightarrow q$, we can, by comparison with

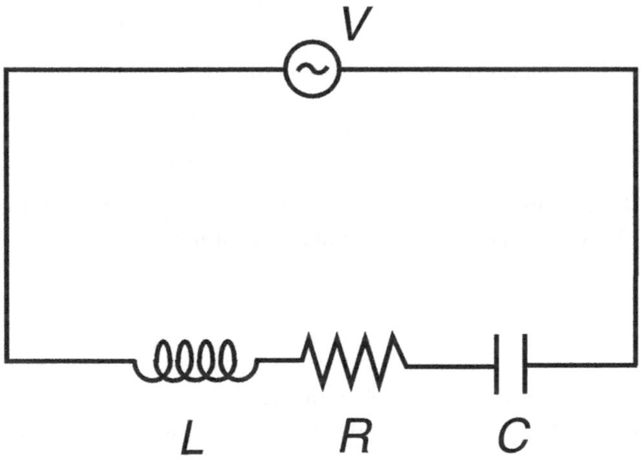

Figure 3.17 LCR circuit.

[5]For this you will need to assume a solution of the form $\tilde{x}(t) = \tilde{A}_0 \exp(i\mu t)$ and show that $\mu = i\gamma/2 \pm [\omega_0^2 - (\gamma/2)^2]^{1/2}$.

Equation (3.43), immediately write the general solution of Equation (3.54) as

$$q(t) = Q_0 e^{-Rt/2L} \cos(\omega t + \phi_0) + Q\cos(\omega_d t + \phi) , \tag{3.55}$$

where Q_0 and ϕ_0 are arbitrary constants, $\omega = \left[\omega_0^2 - (\gamma/2)^2\right]^{1/2}$, and

$$\omega_0 = \left(\frac{1}{LC}\right)^{1/2} , \quad \gamma = \frac{R}{L} , \tag{3.56}$$

$$Q = V \Big/ \sqrt{\left(1/C - L\omega_d^2\right)^2 + R^2\omega_d^2} , \tag{3.57}$$

$$\phi = \tan^{-1}\left[\frac{-R\omega_d}{1/C - L\omega_d^2}\right] . \tag{3.58}$$

All the previous comments concerning resonance apply here also. Amplification at resonance of electrical signals in an LCR circuit is the basis of how a radio set is tuned.

Is there any deep significance that two disparate areas of physics, namely mechanics and electronics, are brought together in forced oscillator theory? Some speak here of the 'unity of physics' without really explaining what they mean. Inductance certainly can be regarded as a kind of electromagnetic inertia, and resistance is definitely akin to damping. However, in my view the significance is superficial rather than profound. It happens that quantities sharing similar properties come together in the same equation, which of course then share the same mathematical solution.

3.12 Power in Forced Oscillations

As we pump away providing the $F_{\text{ext}}(t)$ in the model forced oscillator system, it would be interesting to know how much our efforts are reflected in the movement of the mass. This can be quantified by calculating the power transferred by the driving force *to* the mass as

$$P(t) = F_{\text{ext}}(t) \times \dot{x}(t) , \tag{3.59}$$

which, as indicated, will depend on time. If $P(t)$ is calculated after the transient has died away we can use Equation (3.33) for $x(t)$ and obtain

$$P(t) = -\omega_d AF\cos(\omega_d t) \times \sin(\omega_d t + \phi) ,$$

or, using trigonometry,

$$P(t) = -\frac{\omega_d AF}{2}\left[\sin(2\omega_d t + \phi) + \sin(\phi)\right] . \tag{3.60}$$

Normally we are interested in knowing the power transferred on timescales much greater than T, the oscillator cycle time – the electricity company does not send me a bill every 20ms! – so an appropriate quantity to calculate is the **cycle averaged power**,

$$< P(t) > \equiv \frac{1}{T}\int_t^{t+T} P(t)dt , \tag{3.61}$$

where $T = 2\pi/\omega_d$. Inserting Equation (3.60) into the above integral, the first term cycle averages to zero leaving

$$< P(t) >\equiv -\frac{\omega AF}{2}\sin(\phi)\ . \tag{3.62}$$

The cycle averaged power is always positive since $-\pi \leq \phi \leq 0$, and is maximised for $\phi = -\pi/2$, when the force and the displacement are said to be in **quadrature**. From Equation (3.39), $\phi = -\pi/2$ corresponds to $\omega_d = \omega_0$, which for a lightly damped system is the resonance frequency at which the amplitude is maximised. Utilising Equation (3.41) for the maximum amplitude in Equation (3.62), gives for the maximum possible power transferable by the external force

$$P_{\max} = \frac{1}{2}\frac{F^2}{\gamma m} = \frac{F^2}{2b}\ . \tag{3.63}$$

3.13 Coupled Oscillations

A forced oscillator system is a 'master-slave' system in that the external force compels a mass to move, without the mass itself exerting any influence on whatever is providing the force. We are now going to analyse a system where this asymmetry is removed and the 'driving force' becomes part of the system itself. The model system consists of two masses connected by a spring undergoing mutual harmonic motion as shown in Figure 3.18. As well as occurring frequently in mechanics, coupled oscillator systems are a paradigm for other situations in physics where oscillatory effects are brought into proximity.

Two identical masses of mass m are each attached by a spring of constant k to a rigid support and connected to each other by another spring of constant s. The displacements of the masses from their equilibrium positions are x and y. Although we could devise many other variations of this system, the masses might be unequal for example, the resultant motion for the chosen parameters will illustrate the main features.

For this two-mass problem there is an equation of motion associated with each mass. The mass on the left is pulled back towards its support with a force $-kx$ and forwards towards the other mass with a force $s(y - x)$ so that its equation of motion is

$$m\ddot{x} = -kx + s(y - x)\ . \tag{3.64}$$

Similar reasoning for the mass on the right leads to the equation of motion

$$m\ddot{y} = -s(y - x) - ky\ . \tag{3.65}$$

Notice particularly that unlike the forced oscillator model considered previously, there is no explicit time-dependence in Equations (3.64) and (3.65). A dynamical system in which there is no explicit time-dependence is referred to as **autonomous**, reflecting the fact that it can evolve 'by itself', without external influence.

Dividing both equations of motion by m and introducing the natural frequency $\omega_0 = (k/m)^{1/2}$, we obtain

$$\ddot{x} + \omega_0^2 x - \frac{s}{m}(y - x) = 0\ , \tag{3.66}$$

$$\ddot{y} + \omega_0^2 y + \frac{s}{m}(y - x) = 0\ . \tag{3.67}$$

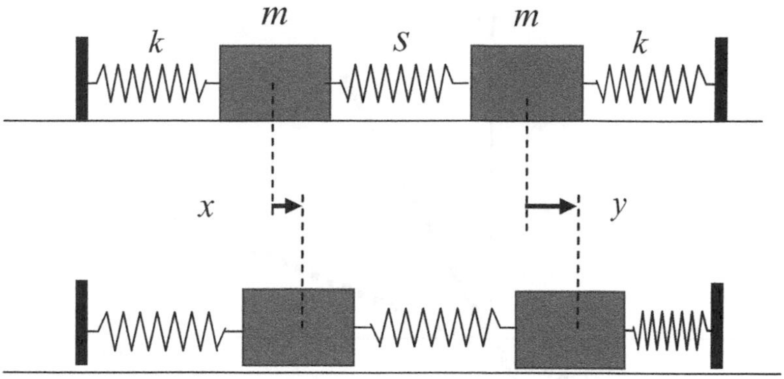

Figure 3.18 Model coupled oscillator system.

We want to find solutions to these equations which are, by the way, *homogeneous*, in the terminology of Section 3.3. There are many possibilities, but we will try to find solutions where the two masses oscillate harmonically at the *same* frequency, ω. In this case the system is said to oscillate in a **normal mode**, and we can write for the coordinates

$$\ddot{x} = -\omega^2 x \text{ and } \ddot{y} = -\omega^2 y \,. \tag{3.68}$$

Under these circumstances the equations of motion become

$$-\omega^2 x + \omega_0^2 x - \frac{s}{m}(y - x) = 0 \,, \tag{3.69}$$

$$-\omega^2 y + \omega_0^2 y + \frac{s}{m}(y - x) = 0 \,. \tag{3.70}$$

We have a pair of linear simultaneous equations. Since the right-hand sides of both equations are zero, then both equations represent lines going through the origin (see Figure 3.19). So the solution would appear to be $x = y = 0$ meaning that both masses just sit there at the origin, which is not very interesting. Is there any way out? Certainly if the two lines coincide then we move from one trivial solution to an infinite number of solutions! Writing the pair of linear simultaneous equations in matrix form

$$\begin{pmatrix} a & b \\ c & d \end{pmatrix} \begin{pmatrix} x \\ y \end{pmatrix} = 0 \,, \tag{3.71}$$

where a, b, c and d are the coefficients appearing in Equations (3.69) and (3.70); then substituting the first equation into the second gives

$$cx + d\left(-\frac{a}{b}\right) x = 0 \,,$$

or the condition that

$$ad - bc = 0 \,. \tag{3.72}$$

If this condition is fulfilled then *any* value of x is a valid solution, and y is then prescribed by $y = (-a/b)x$. The astute reader will recognise the condition given by Equation (3.72)

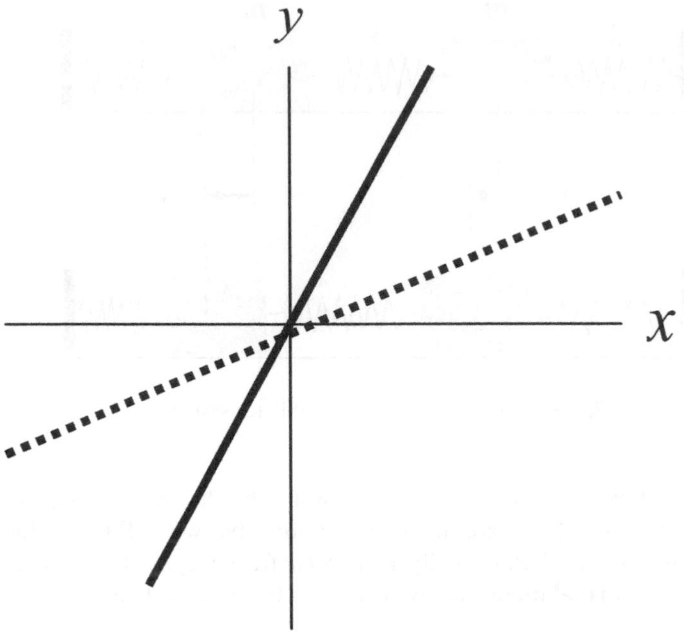

Figure 3.19 The straight lines represented by Equations (3.69) and (3.70).

as the vanishing of the determinant of coefficients of the matrix in Equation (3.71) – see Section A.5. This is a general result in linear algebra. The vanishing of the determinant of coefficients is a necessary and sufficient condition for the existence of nontrivial solutions in a homogeneous linear system.

Applying these considerations to Equations (3.69) and (3.70) results in the following quadratic equation for ω^2:

$$\omega^4 - 2\left(\omega_0^2 + s/m\right)\omega^2 + \left(\omega_0^2 + s/m\right)^2 - \left(s/m\right)^2 = 0\,. \tag{3.73}$$

The solutions of this equation are

$$\omega_1 = \omega_0 \text{ or } \omega_2 = \sqrt{\omega_0^2 + 2\frac{s}{m}} \tag{3.74}$$

and we have calculated the normal mode frequencies. As usual we can ignore negative roots associated with Equations (3.74) as they give identical information.

The next task is to see to which oscillation patterns these normal mode frequencies correspond. Taking $\omega_1 = \omega_0$ first, we can substitute back into either Equation (3.69) or (3.70) and find that for this mode

$$x = y\,, \tag{3.75}$$

that is both masses oscillate harmonically *in phase* with frequency ω_0. Harmonic motion for this mode may therefore be characterised as

$$x = A\cos(\omega_1 t + \phi_1) = y\,, \tag{3.76}$$

where A and ϕ_1 are determined by the initial conditions.

You can prove for yourself that substituting the expression for ω_2 back into either Equation (3.69) or Equation (3.70) results in the oscillation pattern

$$x = -y \; , \tag{3.77}$$

which says that in this mode the masses oscillate at frequency ω_2 in *anti-phase*. Note that the anti-phase oscillation frequency is greater than the in-phase frequency. Harmonic motion for the anti-phase mode is therefore characterised by

$$x = B\cos(\omega_2 t + \phi_2) = -y \; , \tag{3.78}$$

where B and ϕ_2 are two further arbitrary constants. According to Propery 2 in Section 3.3 each second-order differential equation of Equations (3.66) and (3.67) requires two arbitrary constant for its general solution, totalling four arbitrary constants for the whole system. Between them Equations (3.76) amd (3.78) have precisely this number, and their linear combination must therefore represent the general solution:

$$x = A\cos(\omega_1 t + \phi_1) + B\cos(\omega_2 t + \phi_2) \; , \tag{3.79}$$

$$y = A\cos(\omega_1 t + \phi_1) - B\cos(\omega_2 t + \phi_2) \; . \tag{3.80}$$

Now that we have the general solution we can consider a few special cases. If both masses are initially drawn aside by an amount d and released from rest then $A = d$ and $B = 0$. Just the in-phase mode is excited. The phase constant ϕ_1 merely sets the timing for this mode's excitation, whilst ϕ_2 is arbitrary and irrelevant. Figure 3.20 shows the subsequent evolution when the intermediate spring is somewhat weaker than the outer springs ($s = 0.1k$). The in-phase mode is sustained for all time, and the anti-phase mode remains dormant.

If instead we have initially $x = -y = d$, then $A = 0$ and $B = d$, and just the anti-phase mode is excited, as shown in Figure 3.21.

The normal mode oscillations are thus seen to represent special patterns of oscillations which persist for all time. A more interesting case is when one mass is drawn aside and released from rest so that initially, $x = a,\ y = 0$ and $\dot{x} = \dot{y} = 0$. You can show from Equations (3.79) and (3.80) that this results in $A = B = a/2$ and $\phi_1 = \phi_2 = 0$, so that the subsequent evolution is described by

$$x = \frac{a}{2}\left[\cos(\omega_1 t) + \cos(\omega_2 t)\right] \; , \tag{3.81}$$

$$y = \frac{a}{2}\left[\cos(\omega_1 t) - \cos(\omega_2 t)\right] \; . \tag{3.82}$$

In this case *both* modes are present in equal measure. Simple trigonometry allows Equations (3.81) and (3.82) to be recast as

$$x = a\cos(\bar{\omega}t)\cos(\delta\omega t) \; , \tag{3.83}$$

$$y = a\sin(\bar{\omega}t)\sin(\delta\omega t) \; , \tag{3.84}$$

where $\bar{\omega} \equiv (\omega_1 + \omega_2)/2$ is the average frequency and $\delta\omega \equiv (\omega_2 - \omega_1)/2$. For the same mass and spring constant values for Figures 3.20 and 3.21 the evolution is now illustrated in Figure

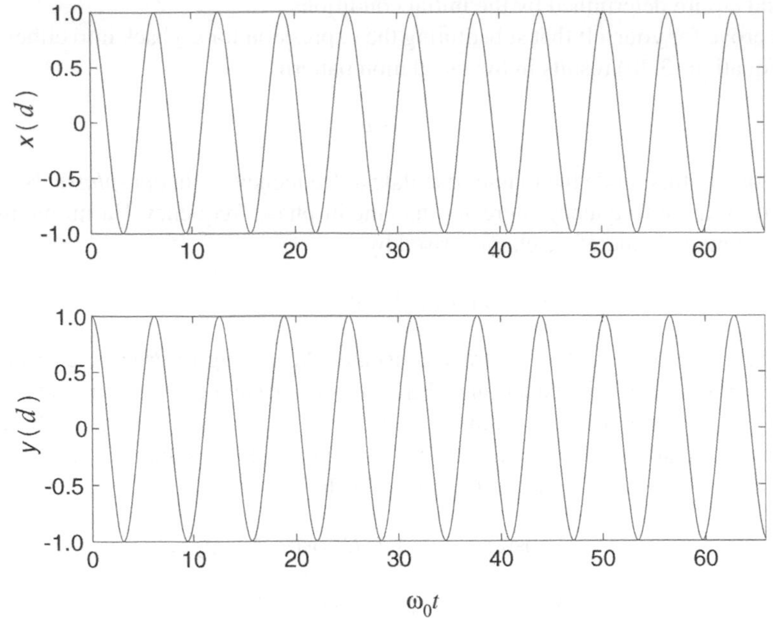

Figure 3.20 Evolution of the in-phase mode in the coupled oscillator system when initially $x = y = d$.

3.22. The system now exhibits the phenomenon of **beats** in which both masses oscillate at the average frequency $\bar{\omega}$, but energy is exchanged between the masses at the **beat frequency**, $\delta\omega$.

Amongst the nonmechanical situations to which coupled oscillator theory applies we mention the phenomenon of light passing from one optical fibre to a parallel neighbouring fibre as shown in Figure 3.23. It turns out that the propagation of light along the parallel fibres is described by very similar equations to the coupled oscillator system considered in this section. Light initially present in one fibre excites light in the fibre next door in a manner very similar to the excitation of the neighbouring mass in the coupled oscillator system. Beating in this case consists of the light zigzagging to and fro between the two fibres at a rate given by the so-called coupling length. If the fibres are separated after one coupling length then the light has been transferred from one fibre to the other. This is a useful device in integrated optical systems.

In analysing the coupled oscillator system we have made an initial step towards the general situation of oscillations of *extended* systems. Many masses connected by springs in a lattice constitutes a model for atomic vibrations in solids. In that case many more normal modes will emerge, because we are then dealing with huge numbers of masses, although the principles developed in this section still apply. Taking the continuum limit of multi-mass systems constitutes the study of **waves** supported by continuous media, but that is as far as we wish to take the matter.

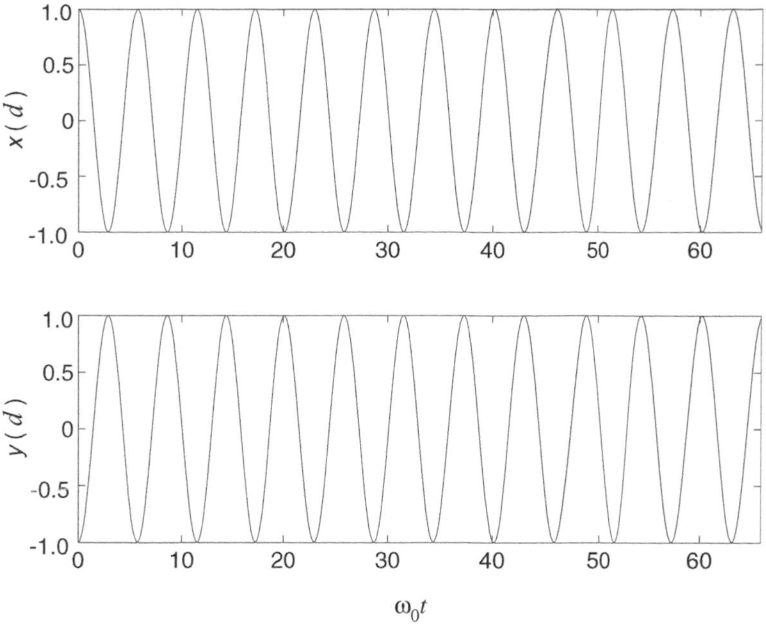

Figure 3.21 Evolution of the anti-phase mode in the coupled oscillator system when initially $x = -y = d$.

3.14 Summary

Section 3.9. The motion of the forced, damped oscillating system of Figure 3.9 may be represented by

$$\ddot{x} + \gamma\dot{x} + \omega_0^2 x = \frac{F\cos(\omega_d t)}{m} \tag{3.32}$$

where $\gamma = b/m$ represents the damping and $\omega_0^2 = k/m$ is proportional to the restoring force provided by a spring. The general solution of Equation (3.32) is given by

$$x(t) = \underbrace{A_0 e^{-\gamma t/2}\cos(\omega t + \phi_0)}_{\text{Transient}} + \underbrace{A\cos(\omega_d t + \phi)}_{\text{Steady-state}} \tag{3.43}$$

where $q = \gamma/2$ and

$$\omega^2 = \omega_0^2 - \gamma^2/4\,. \tag{3.19}$$

The oscillation amplitude A_0, and phase angle ϕ_0, are determined by the initial conditions, specified via $x(0)$ and $\dot{x}(0)$.

Special cases:

1. $F = 0$ and $\gamma = 0$ (**Section 3.2** and **Section 3.4**). The mass oscillates at the natural angular frequency $\omega_0 = (k/m)^{1/2}$ with constant amplitude and phase angle

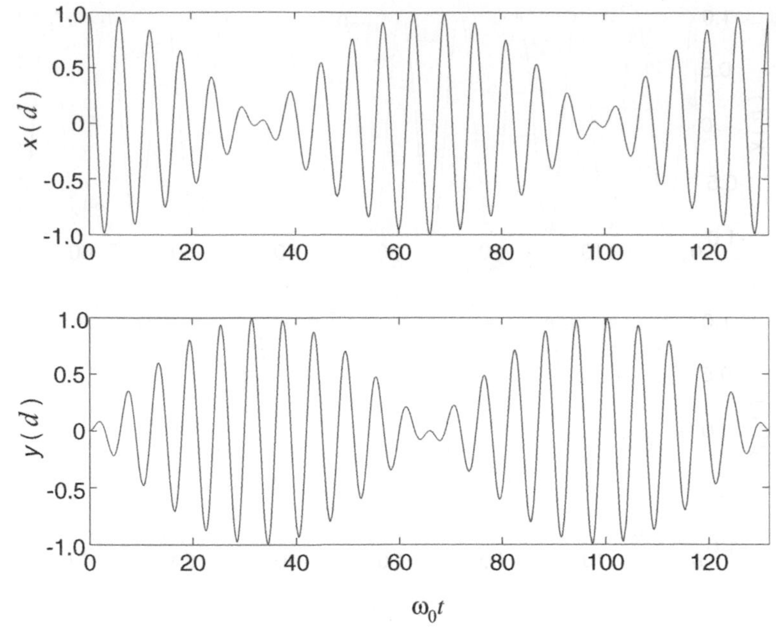

Figure 3.22 Evolution of x and y in the coupled oscillator system when initially $x = d$ and $y = 0$.

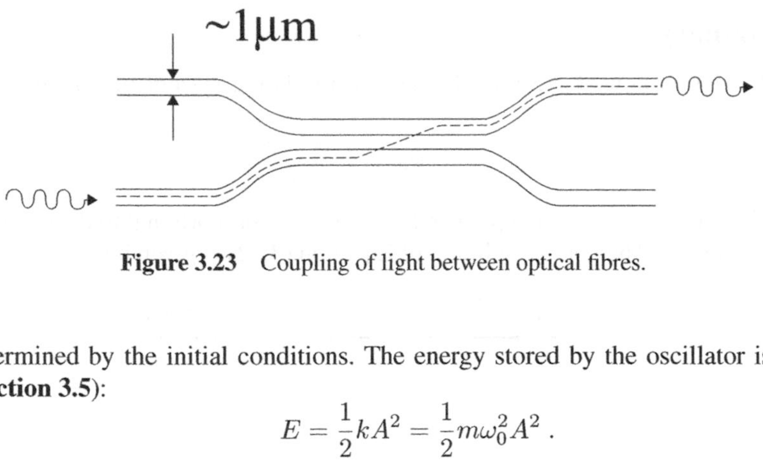

Figure 3.23 Coupling of light between optical fibres.

determined by the initial conditions. The energy stored by the oscillator is given by (**Section 3.5**):

$$E = \frac{1}{2}kA^2 = \frac{1}{2}m\omega_0^2 A^2 . \tag{3.10}$$

2. $F = 0$ and $\gamma \neq 0$.

   (a) Light damping $\omega_0 > \gamma/2$ (**Sections 3.6**) The mass oscillates at a frequency $\omega = (\omega_0^2 - \gamma^2/4)^{1/2}$. The ratio $Q = \omega_0/\gamma$ measures the loss of energy per cycle (**Section 3.7**).

   (b) Critical damping $\omega_0 = \gamma/2$ (**Section 3.8**). The mass does not oscillate, but it approaches the equilibrium state in the shortest possible time.

(c) Heavy damping $\omega_0 < \gamma/2$ (**Section 3.8**). Solutions are exponential in character, the system relaxing to equilibrium, first quickly and then slowly, without oscillation.

3. $F \neq 0$ and $\gamma \neq 0$ (**Section 3.9**).

Initially both parts of Equation (3.43) are relevant. The behaviour can be quite violent initially. Whatever the initial conditions, the long-term behaviour is given by the steady-state after the transient has died away. The amplitude, A, and the phase, ϕ, are determined as

$$A = \frac{F}{m} \Big/ \sqrt{\left(\omega_0^2 - \omega_d^2\right)^2 + \gamma^2\omega_d^2}\,, \tag{3.52}$$

and

$$\tan\phi = \frac{-\gamma\omega_d}{\omega_0^2 - \omega_d^2}\,. \tag{3.53}$$

The cycle-averaged power in a forced oscillator is (**Section 3.12**)

$$< P(t) > \equiv -\frac{\omega A F}{2}\sin(\phi). \tag{3.62}$$

Section 3.13. **Coupled oscillations** between two identical masses are described by the equations of motion

$$\ddot{x} + \omega_0^2 x - \frac{s}{m}(y - x) = 0\,, \tag{3.66}$$

$$\ddot{y} + \omega_0^2 y + \frac{s}{m}(y - x) = 0\,. \tag{3.67}$$

The general solution of this coupled system is

$$x = A\cos(\omega_1 t + \phi_1) + B\cos(\omega_2 t + \phi_2)\,, \tag{3.79}$$

$$y = A\cos(\omega_1 t + \phi_1) - B\cos(\omega_2 t + \phi_2)\,. \tag{3.80}$$

where

$$\omega_1 = \omega_0 \text{ and } \omega_2 = \sqrt{\omega_0^2 + 2\frac{s}{m}}\,. \tag{3.74}$$

A, B, ϕ_1 and ϕ_2 are determined by the initial conditions. The motion is therefore a combination of two modes, one where the masses oscillate in phase at ω_1 and one where they oscillate in anti-phase at ω_2. Generally energy is continually transferred between one mass and the other at a rate given by the beat frequency $\delta\omega \equiv (\omega_2 - \omega_1)/2$.

3.15 Problems

3.1 A vibrating object undergoes four complete cycles in 2 s. Calculate the angular frequency, the frequency and the period of the motion.

3.2 A simple harmonic oscillator uses a 0.4 kg mass and a spring of unknown spring constant. The period is 0.2 s. What is the spring constant?

3.3 An object of mass 0.3 kg undergoes SHM at the end of a horizontal spring (constant $k = 270$ N m^{-1}). When the object is 1 cm from its equilibrium position it has a speed of 0.3 m s^{-1}. What is (i) the amplitude of the motion, (ii) the total energy and (iii) the maximum speed attained?

3.4 A pendulum has a period of 2 s when it oscillates in a vacuum. When in air, its amplitude is reduced by a factor of two after 10 s.

(a) By what factor is the energy reduced after 10 s?

(b) By what factor is the energy reduced after 30 s?

(c) Calculate the period of the pendulum in air.

3.5 Show that the velocity in an unforced damped harmonic oscillator is a maximum when $\omega = \omega_0$.

3.6 Using the correspondence between electrical $\leftrightarrow$ mechanical oscillators calculate the frequency at which the current is a maximum in an electrical circuit which includes a 1μF capacitor and a 10 mH inductor in series with an AC voltage source. (The farad (F) and henry (H) are the SI units of capacitance and inductance, respectively.)

3.7 A cuckoo clock has a pendulum of length 20 cm with a 20 g mass at the end. The amplitude of the oscillation is maintained at 0.2 rad by a weight of mass 200 g which falls 2 m during 24 h. Estimate the Q of the system. How long would the clock run if powered by a (typical) 10 kJ battery, instead of the weight?

3.8 A mass m hangs vertically at the end of a spring (see Figure 3.24). If the spring constant is k, obtain an expression (in terms of m, k and g) for the static equilibrium extension, x_0.

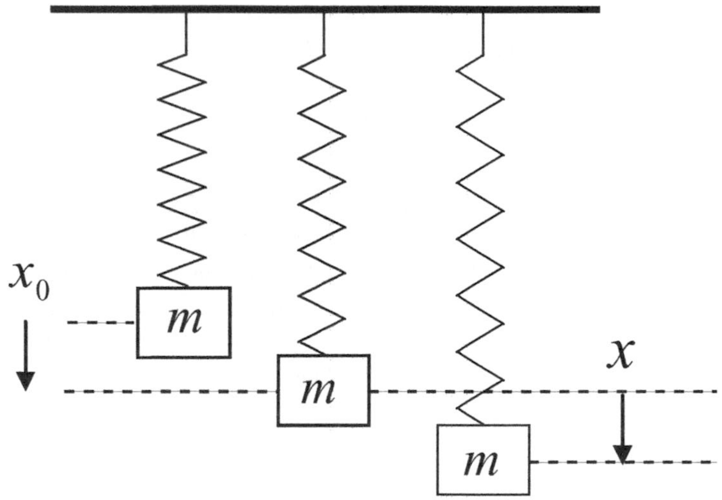

Figure 3.24 Figure for Problem 3.8.

(a) If x is a further displacement (measured downwards from the equilibrium position) show that the resultant force on the mass is given by $-kx$. Hence show that the equation of the subsequent motion can be written

$$\ddot{x} + \omega_0^2 x = 0, \quad \text{where} \quad \omega_0^2 = k/m\,.$$

(b) Verify that $x = A\cos(\omega_0 t + \phi)$ is a solution.

At time $t = 0$, while the mass is stationary at the equilibrium position, it receives an impulse in the upward direction. You may assume that: $m = 0.5$ kg, $k = 12.5\ \text{Nm}^{-1}$, impulse given to mass $= -2$ N s (negative sign signifies the upward direction!).

(c) What are the initial values of x and $\dot{x}$?

(d) Using these initial conditions, calculate ϕ and A.

(e) What is the period of the oscillations?

(f) When air resistance is taken into account there is a damping force given by $F = -\gamma m v$, where v is the velocity and γ is the damping constant. The amplitude of the oscillation is observed to decrease by a factor of two after 100 s. By what factor has the energy of the oscillator decreased after 100 s?

(g) Calculate the value of the damping constant, γ.

(h) Calculate the Q value for the oscillator.

(i) Calculate the *fractional* increase in the period compared with the undamped SHM period. Your calculator is probably able to perform this calculation, but as an exercise, don't use it. *Hint*: use a binomial approximation for the expression $\omega^2 = \omega_0^2 - (\gamma/2)^2$ relating the undamped angular frequency, ω_0, to the damped frequency, ω.

3.9 A child of mass 25 kg is set swinging on a swing with an amplitude of 0.1 rad. The effective length l of the swing is 2.5 m.

(a) Using a small angle approximation, show that the lateral displacement x of the swing satisfies the equation $\ddot{x} + (g/l)x \approx 0$.

(b) Calculate the period with which the child oscillates.

(c) Calculate the energy stored in the system.

Air resistance introduces an additional term $(b/m)\dot{x}$ to the left-hand side of the equation in part (a), where $b = 0.3\,\text{kg}\,\text{s}^{-1}$.

(d) Calculate the Q of the system.

(e) Calculate the time for the child's oscillations to decay to half their original value.

(f) Calculate the frequency at which the parent must push to increase the amplitude with minimum effort, giving your answer to two decimal places.

(g) With what phase should the parent push relative to the swing?

3.10 A crucial bolt in a bridge snaps, allowing the section it secured (of mass $m = 3 \times 10^4$ kg) to move in response to forces with a restoring force of 1.13×10^7 N per metre of induced displacement. The system is lightly damped (with $b = 1.2 \times 10^3\ \text{kg s}^{-1}$)

and if the amplitude of any induced movement exceeds 1.0 m, the bridge will collapse. A troop of 40 soldiers each weighing approximately 75 kg march over the bridge in step. The force each soldier exerts on the bridge as he walks may be modelled as sinusoidal, and the stride length of each soldier is approximately 1m. Calculate the marching speed most likely to cause damage to the bridge, and estimate whether or not the bridge will collapse.

3.11 A van of mass 10^3 kg, when stationary on level ground, compresses its suspension springs by 10 cm (neglect the mass of the wheels, and anything else which is not supported by the suspension).

(a) Calculate the *effective* spring constant, k (that is that resulting from considering the mass of the truck compressing a single spring by 10 cm).

(b) Calculate the *undamped* oscillation period.

(c) Shock absorbers (dampers) are fitted to provide *critical* damping. Calculate the resistive force constant, b.

(d) The van is now loaded with an *additional* 3 x 10^3 kg. Calculate the new values, ω_0' and γ' of the free oscillation angular frequency and the damping constant. Show that the system is now *under-damped*.

(e) Calculate the oscillation period of the loaded van.

(f) The loaded van drives at a constant speed along a road with a series of speed restricting bumps (which provide a vertical periodic external force). Assume that the road surface can be approximated by a cosine function with 20 m separating the peaks. With what speed is the van traveling when resonance occurs between the bumps and the suspension?

(g) The comfort of the ride is improved when oscillations of the height of the van above the mean road surface are reduced. Is this best achieved by driving faster or slower than the resonance speed? (*Hint*: consider the phase difference between the van's displacement from equilibrium and the vertical profile of the road surface.)

3.12 Two identical pendulums are connected together as shown in Figure 3.25. Each has a length l and a mass m. This problem calculates the frequencies of the normal modes for *small amplitude* oscillations of this coupled system. Use small angle approximations: so neglect all terms of order θ^2 and higher in the following.

(a) Express the displacements from equilibrium x_1 and x_2, in terms of the angles θ and ϕ, assuming θ and ϕ are *small*.

(b) Show that (to the same order of approximation) the tensions T_1 and T_2 in the two strings can be written: $T_1 = 2mg$ and $T_2 = mg$.

(c) Hence write down the equations of motion for the masses and show that they reduce to:

$$\ddot{\theta} = -(g/l)(2\theta - \phi) ,$$

$$\ddot{\theta} + \ddot{\phi} = -(g/l)(\phi) .$$

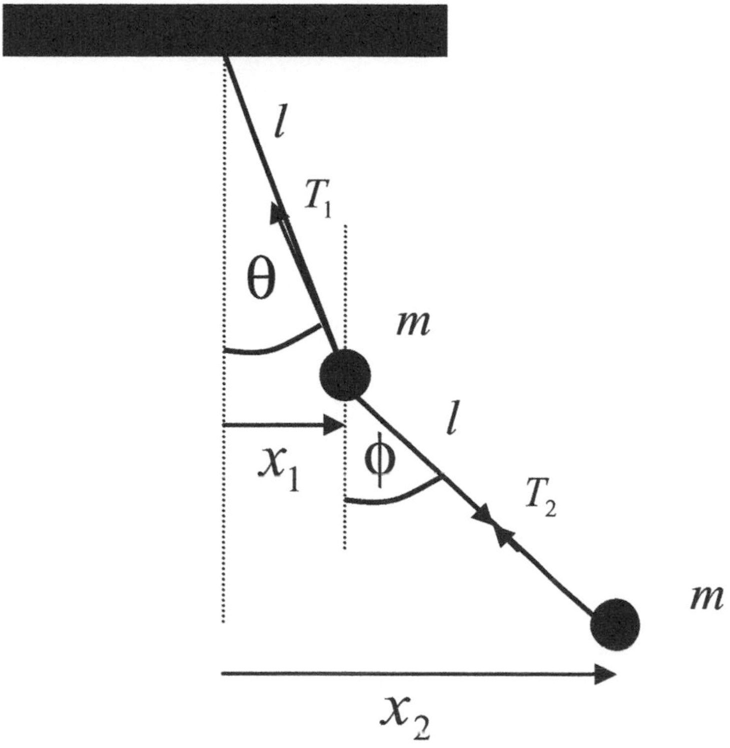

Figure 3.25 Coupled pendulums for Problem 3.12.

(d) Write down the condition for normal mode oscillations (of angular frequency ω) of the system and show that this leads to a pair of simultaneous algebraic equations:

$$\left(-\omega^2 + \frac{2g}{l}\right)\theta - \frac{g}{l}\phi = 0\,,$$

$$\left(-\omega^2\right)\theta + \left(-\omega^2 + \frac{g}{l}\right)\phi = 0\,.$$

(e) Hence show that the normal mode frequencies are given by

$$\omega_{1,2}^2 = (2 \pm \sqrt{2})(g/l) \quad .$$

(f) Calculate ω_1 and ω_2 for $l = 0.5$ m and $g = 10$ m s^{-2} and compare with the value of ω for a simple pendulum of length $l = 0.5$ m.

(g) By substituting ω_1 and ω_2 back into the above equations obtain relations between θ and ϕ. This tells you what the modes look like. Sketch the time dependence of θ and ϕ for each of the two modes.

Figure 2.15 Coupled pendulums for Problem 2.12.

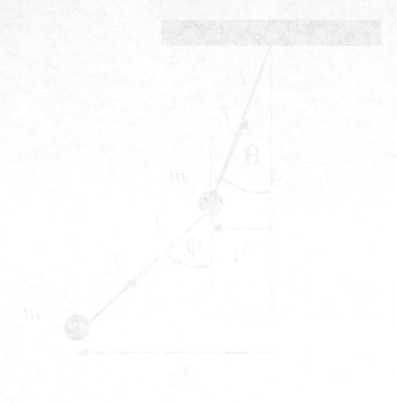

4

Two-body Dynamics

4.1 Rationale

Sometimes it is useful to know how a collection of bodies behave 'as a whole', and to deduce general properties of the system. For example, a container filled with gas contains billions of molecules. We could analyse the system by writing down Newton's laws for all the molecules, but this would be far too tedious. We are normally interested in general properties, such as the pressure, temperature, etc. These macroscopic features are the domain of thermodynamics and statistical physics. Here, as a first step, we are going to see how two bodies can be described collectively. To give some notion of what the 'collective' behaviour of two bodies might mean, imagine two masses connected by a fairly rigid spring, the whole being thrown across a room. Those close to the dumb-bell-like composite body will see the two masses oscillating between each other as they fly through the air together. Those further away, however, will have a less precise picture. They will be unable to resolve the detailed oscillations, and will only see the parabolic trajectory of the whole structure. Our initial aim, therefore, is to make this dissection into 'internal' motion and 'general' motion more precise.

4.2 Centre of Mass

Keeping the dumb-bell paradigm in mind, the equations of motion of the two masses are

$$m_1\ddot{\mathbf{r}}_1 = \left(\mathbf{F}_{12} + \mathbf{F}_1^{\text{Ext}}\right) , \tag{4.1}$$

and

$$m_2\ddot{\mathbf{r}}_2 = \left(\mathbf{F}_{21} + \mathbf{F}_2^{\text{Ext}}\right) , \tag{4.2}$$

where $\mathbf{F}_{12}$ is the force on particle 1 due to particle 2, and $\mathbf{F}_1^{\text{Ext}}$ is the *external* force acting on particle 1 (e.g. gravity). Similarly for particle 2. Adding these two equations

$$m_1\ddot{\mathbf{r}}_1 + m_2\ddot{\mathbf{r}}_2 = \left(\mathbf{F}_1^{\text{Ext}} + \mathbf{F}_2^{\text{Ext}}\right) = \mathbf{F}_{\text{tot}}^{\text{Ext}} , \tag{4.3}$$

where $\mathbf{F}_{\text{tot}}^{\text{Ext}}$ is the *total* external force acting on the system, and we have used the fact that $\mathbf{F}_{12} = -\mathbf{F}_{21}$ by Newton's third law. Now defining a vector $\mathbf{R}$, according to

$$\mathbf{R} = \frac{m_1\mathbf{r}_1 + m_2\mathbf{r}_2}{m_1 + m_2} , \tag{4.4}$$

Classical Mechanics – Second Edition From Newton to Einstein: A Modern Introduction Martin W. McCall

Equation (4.3) becomes

$$M\ddot{\mathbf{R}} = \mathbf{F}_{\text{tot}}^{\text{Ext}} , \tag{4.5}$$

where $M = m_1 + m_2$ is the total mass. The system as a whole may therefore be described as moving according to Newton's second law with total mass M and position vector $\mathbf{R}$, in response to the total external force $\mathbf{F}_{\text{tot}}^{\text{Ext}}$. The vector $\mathbf{R}$ is called the **centre of mass** of the two-body system. Evidently if $\mathbf{F}_{\text{tot}}^{\text{Ext}} = \mathbf{0}$ then

$$M\dot{\mathbf{R}} = \text{constant} , \tag{4.6}$$

and the system momentum is conserved, for then no net external force acts and the the system is isolated.

By way of a simple example, consider two unconnected bodies falling vertically under gravity. Then the individual equations of motion, $m_1\ddot{z}_1 = -m_1 g$ and $m_2\ddot{z}_2 = -m_2 g$, sum to give

$$M\ddot{Z} = -m_1 g - m_2 g = -Mg , \tag{4.7}$$

where $Z \equiv (m_1 z_1 + m_2 z_2)/(m_1 + m_2)$ is the centre-of-mass coordinate. The centre of mass is thus also seen to be in freefall with the same acceleration due to gravity.

4.3 Internal Motion: Reduced Mass

Now what about the internal motion of our spring-connected dumb-bell? If gravity is the external force then $\mathbf{F}_1^{\text{Ext}}/m_1 = \mathbf{F}_2^{\text{Ext}}/m_2 = \mathbf{g}$, the acceleration due to gravity, and subtracting Equation (4.1) from Equation (4.2) we obtain

$$\ddot{\mathbf{r}}_2 - \ddot{\mathbf{r}}_1 = \left(\frac{1}{m_1} + \frac{1}{m_2}\right)\mathbf{F}^{\text{int}} = \mu^{-1}\mathbf{F}^{\text{int}} , \tag{4.8}$$

where we have defined $\mathbf{F}^{\text{int}} \equiv \mathbf{F}_{21} = -\mathbf{F}_{12}$ and μ, the so-called **reduced mass**, according to

$$\mu \equiv \left(\frac{1}{m_1} + \frac{1}{m_2}\right)^{-1} . \tag{4.9}$$

Introducing the **relative coordinate** $\mathbf{r} \equiv \mathbf{r}_2 - \mathbf{r}_1$, the equation of motion for the internal motion is

$$\mu\ddot{\mathbf{r}} = \mathbf{F}^{\text{int}} . \tag{4.10}$$

The reduced mass is so-called because it is always less than either m_1 or m_2 separately. For the example of two masses connected by a spring, the internal force is just the extension times the spring constant, so that in one dimension

$$\mu\ddot{x} = -kx . \tag{4.11}$$

The masses execute SHM with angular frequency given by

$$\omega_0 = \left(\frac{k}{\mu}\right)^{1/2} = \left[k\left(\frac{m_1 + m_2}{m_1 m_2}\right)\right]^{1/2} . \tag{4.12}$$

If the external force is such that the external acceleration imparted to each body is not equal, then additional terms would appear on the right of Equation (4.8) and the motion is then not so simple. It is also interesting to note that, unlike the centre of mass, the reduced mass is not naturally extended to many bodies, or to continuous distributions of matter; it is a purely two-body concept.

Example: Calculate the ratio of the natural vibration frequencies of the hydrogen molecule H_2 to the hydrogen chloride molecule HCl assuming that each molecule has the same bonding strength, and that the mass of a Cl atom is 35 times the mass of a hydrogen atom.

$$\mu_{\mathrm{H_2}} = \frac{1\times 1}{1+1} = \frac{1}{2} \quad \Rightarrow (\omega_0)^2_{\mathrm{H_2}} = \frac{k}{(1/2)} ,$$

$$\mu_{\mathrm{HCl}} = \frac{1\times 35}{1+35} \approx 1 \Rightarrow (\omega_0)^2_{\mathrm{HCl}} = \left(\frac{k}{1}\right) .$$

So

$$(\omega_0)_{\mathrm{H_2}} = \sqrt{2}\,(\omega_0)_{\mathrm{HCl}} \ .$$

The answer is true for any hydrogen X-ide molecule where X is much heavier than hydrogen.

4.4 Collisions

A **collision** occurs when two bodies interact and the time taken for the interaction is 'small' in the sense that our interest lies in the particles' dynamics for times before and after the interaction occurs, when the particles are 'free'. It is not necessary for the bodies to physically touch for a collision to occur. Coulomb repulsion of like charges would be equally valid. The most general situation is depicted in Figure 4.1, in which two particles approach each other following their respective straight lines in three-dimensional space with respect to some inertial frame. Now already it is possible to simplify the description by moving to an inertial frame in which one axis (x, say) is aligned to the direction along which the particles are approaching each other (see Figure 4.2). We define a one-dimensional collision as one in which the masses still move along the x-axis (as defined above) *after* the collision:

- A one-dimensional collision between two particles is one where the line connecting the particles does not change direction.

It is possible for the particles to scatter obliquely so that after the collision they are not moving along the line of approach. However, whilst applying momentum conservation to such collisions is straightforward, the general analysis of oblique collisions is quite complicated and will not be discussed in this book.

For the two bodies approaching each other and scattering along the x-axis we have, in the notation of Section 4.3

$$m_1\ddot{x}_1 = -F^{\mathrm{int}} , \tag{4.13}$$

$$m_2\ddot{x}_2 = F^{\mathrm{int}} , \tag{4.14}$$

where the superscript 'int' may now stand for either 'internal' as in Section 4.3, or, perhaps better, 'interaction'. Either way, summing Equations (4.13) and (4.14) and integrating we obtain

$$\int_{\mathrm{before}}^{\mathrm{after}} \left[m_1\frac{d\dot{x}_1}{dt} + m_2\frac{d\dot{x}_2}{dt}\right] dt = [m_1\dot{x}_1 + m_2\dot{x}_2]^{\mathrm{after}}_{\mathrm{before}} \ . \tag{4.15}$$

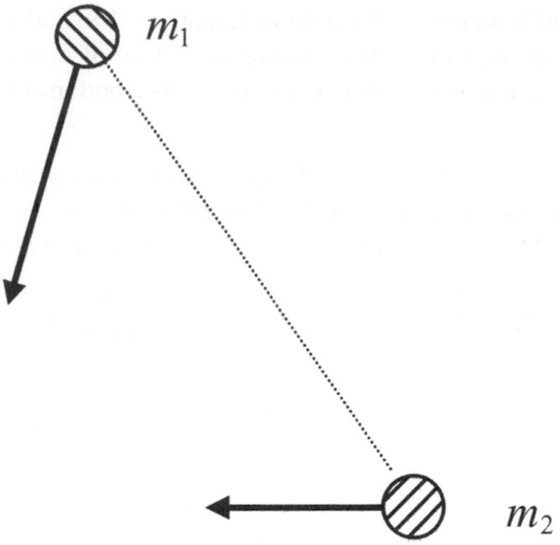

Figure 4.1 Two bodies approaching before a collision.

Figure 4.2 Bodies approaching each other as in Figure 4.1, but viewed from a frame in which the line joining the centre of the bodies is stationary.

If u_1 and u_2 are the velocities before, and v_1 and v_2 are the velocities after, then Equation (4.15) becomes

$$m_1u_1 + m_2u_2 = m_1v_1 + m_2v_2 \, , \tag{4.16}$$

reflecting again momentum conservation for an isolated system. But note that this refers explicitly to the momentum before and after the collision. Of course momentum is also conserved *during* the collision as well (when the velocities will be different from $u_{1,2}$ and $v_{1,2}$), but that is not our interest here.

For isolated systems momentum conservation is 'always' true. In fact, as the particle speeds approach that of light, momentum conservation as expressed in Equation (4.16) is experimentally *false*, but when we study relativity in the next two chapters, momentum will be redefined such that it is still conserved, and reproduces Equation (4.16) in the low velocity limit. We will change the rules of the game! This is not an arbitrary process, as we shall see.

4.5 Elastic Collisions

If F^{int} is a conservative force then the total mechanical energy is conserved:

$$E = K + U^{\text{int}} \, , \tag{4.17}$$

where $U^{\text{int}}(x)$, the interaction potential, related to F^{int} by

$$F^{\text{int}} = -\frac{dU^{\text{int}}}{dx} \, . \tag{4.18}$$

As we emphasised at the beginning of this section, our interest here lies in what happens before and after the particles interact, so we regard the range of x over which $U^{\text{int}}(x)$ is effective as being negligibly small. This is the case, for example, when two billiard balls collide, for then the interaction potential is the electrostatic repulsion occurring when the atoms of one ball try to penetrate into the atoms of the other. This acts over such a short range, and for such a short time, that we regard the collision as 'hard'. For most of the time $U^{\text{int}}(x)$ is constant, so that energy conservation becomes

$$K_{\text{initial}} = \frac{1}{2}m_1u_1^2 + \frac{1}{2}m_2u_2^2 = \frac{1}{2}m_1v_1^2 + \frac{1}{2}m_2v_2^2 = K_{\text{final}} \, . \tag{4.19}$$

Collisions for which kinetic energy is conserved are termed **elastic**.

Example: Given initial velocities, calculate the final velocities of two bodies which collide elastically in one dimension.

Energy conservation, Equation (4.19) , can be rewritten as

$$m_1(v_1^2 - u_1^2) = m_2(u_2^2 - v_1^2) \, . \tag{4.20}$$

Momentum conservation, Equation (4.16), can be rewritten as

$$m_1(v_1 - u_1) = m_2(u_2 - v_2) \, . \tag{4.21}$$

Dividing the last two equations yields

$$(v_1 + u_1) = (u_2 + v_2) \tag{4.22}$$

or

$$(v_2 - v_1) = -(u_2 - u_1) \, , \tag{4.23}$$

that is the **relative velocity** is reversed.

We can now use this information to calculate the final velocities v_1 and v_2 in terms of the initial velocities u_1 and u_2, by solving Equations (4.21) and (4.22) simultaneously. The result is

$$v_1 = \frac{2m_2u_2 + (m_1 - m_2)u_1}{m_1 + m_2} \, , \tag{4.24}$$

and

$$v_2 = \frac{2m_1u_1 + (m_2 - m_1)u_2}{m_1 + m_2} \, . \tag{4.25}$$

For the special case of scattering from a stationary target, $u_2 = 0$ and

$$v_1 = \left(\frac{m_1 - m_2}{m_1 + m_2}\right) u_1 \, , \quad v_2 = \frac{2m_1}{m_1 + m_2} u_1 \, . \tag{4.26}$$

From these equations further special cases emerge:

If:	$m_1 = m_2$: $v_1 = 0$, $v_2 = u_1$	The kinetic energy of the incident body is completely transferred to the stationary body.
	$m_1 > m_2$: v_1 and $v_2 > 0$	Both bodies move along the positive x axis.
	$m_1 < m_2$: $v_1 < 0$, $v_2 > 0$	The incident body 'rebounds'.
	$m_1 \ll m_2$: $v_1 \approx -u_1$, $v_2 \approx 0$	A 'brickwall' rebound; the incident body is reflected from the massive body which remains stationary.

These cases do not present any particularly surprising features, except perhaps the brick wall rebound which apparently violates momentum conservation. In that case the momentum before the collision is $m_1 u_1$, whereas after it is $-m_1 u_1$. The resolution comes from carefully examining the exact equation for momentum conservation which, taking Equation (4.26) into account, becomes

$$m_1 u_1 = m_1 \left(\frac{m_1 - m_2}{m_1 + m_2}\right) u_1 + m_2 \left(\frac{2m_1}{m_1 + m_2}\right) u_1 . \tag{4.27}$$

Now in the limit $m_2 \gg m_1$ this reads

$$m_1 u_1 = -m_1 u_1 + 2m_1 u_1 . \tag{4.28}$$

The recoiling momentum of the massive body is $+2m_1 u_1$. The recoil *velocity* of the massive body is just the momentum divided by the mass, or $2m_1 u_1 / m_2$, and this tends to zero in the limit $m_1 \ll m_2$. Multiplying this zero velocity with a large mass yields the finite *momentum* $+2m_1 u_1$ for the larger mass.

Another special case is when two bodies collide elastically with equal and opposite velocity. With what speed does the lighter mass recoil (see Figure 4.3)? Here $u_1 = -u_2 \equiv u$,

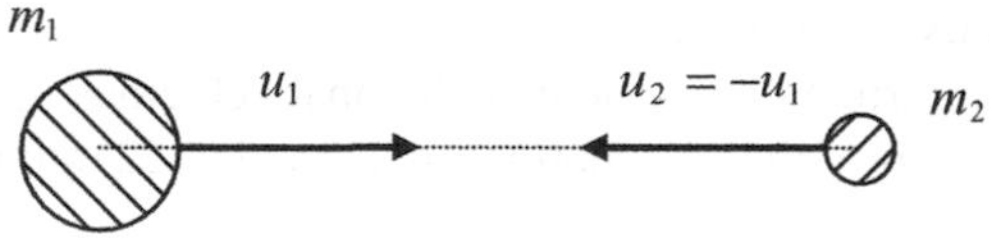

Figure 4.3 Two bodies of unequal mass approaching each other with equal speeds.

so from Equation (4.25)

$$v_2 = \frac{2m_1 u + (m_2 - m_1)(-u)}{m_1 + m_2} = \frac{3m_1 - m_2}{m_1 + m_2} u , \tag{4.29}$$

or

$$v_2 = \left[\frac{3\rho - 1}{\rho + 1}\right] u , \tag{4.30}$$

where $\rho \equiv m_1/m_2$. For equal masses, $\rho = 1$, $v_1 = -u$, $v_2 = u$ and the bodies bounce off each other symmetrically. For $m_1 \gg m_2$, $\rho \gg 1$ and $v_2 \to 3u$. Thus when two bodies approach each other with equal but opposite velocities the maximum recoil speed of the lighter body is three times the speed of the heavier body. During my lectures I give an amusing demonstration of this formula. I take a column through which a (relatively) large ball is fixed, whilst a lighter ball is also threaded onto the column, but is free to move. Both balls are made out of rubber and collide fairly elastically (see Figure 4.4). The column is

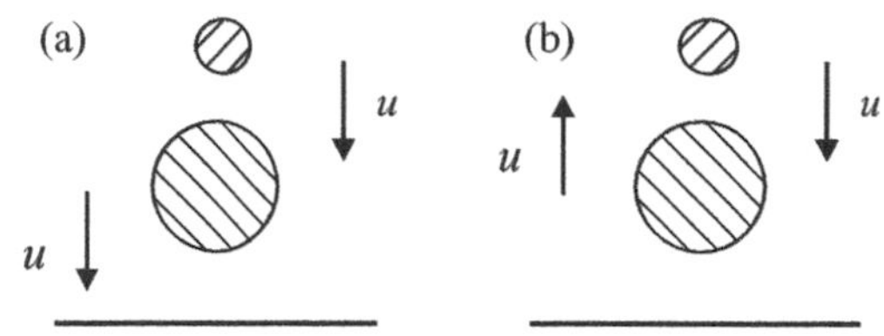

Figure 4.4 Bounce of two bodies of unequal mass falling initially with the same velocity (a) before the heavier ball hits the ground, and (b) after the heavier ball hits the ground.

dropped from a height h (about 1 m), and hits the ground with speed $u = (2gh)^{1/2}$. The lower mass makes contact with the ground, reversing its velocity (='brick wall' collision with the Earth!) so that on impact effectively the two masses collide with equal and opposite velocities. Using Equation (4.30) and the fact that the height, H, attained by the recoiling lighter ball will be proportional to the square of the recoil velocity we have

$$H = \left(\frac{3\rho - 1}{\rho + 1}\right)^2 h \to 9h \tag{4.31}$$

in the limit $\rho \gg 1$. Figure 4.5 shows a plot of the height attained by the lighter mass as a function of the mass ratio ρ, showing particularly the asymptotic behaviour described by Equation (4.31). Even accounting for the fact that in practice the collision is not perfectly elastic, the upper ball generally hits the ceiling of the lecture theatre. Incidentally, this is pretty much how a supernova works. A supernova is initiated when the central material of a massive star implodes. On meeting at the middle the material effectively bounces off itself and explodes with equal and opposite radial velocity. Imploding lighter outer material meets the exploding massive core causing the lighter material to bounce off at much higher speed, resulting in the supernova explosion.

4.6 Inelastic Collisions

Inelastic collisions occur when kinetic energy is no longer conserved so that $K_{\text{final}} < K_{\text{initial}}$. We saw in Section 4.5 that an elastic collision was characterised by the reversal of the relative velocity between the two bodies (see Equation (4.23)), and therefore the maintenance of the relative speed. It is therefore natural to quantify inelastic collisions by the degree to which the relative speed is no longer conserved. This is achieved through the so-called **coefficient of restitution**, ε.

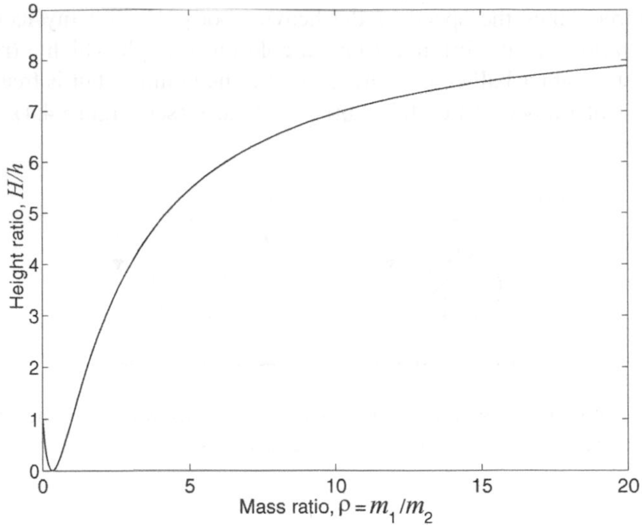

Figure 4.5 Plot of Equation (4.31).

- The coefficient of restitution, ε, in a collision is defined as the ratio of the relative velocity of separation, divided by the relative velocity of approach:

$$v_2 - v_1 = \varepsilon\,(u_1 - u_2)\ . \tag{4.32}$$

Experimentally, ε is roughly constant for given mass pairs. Two extreme limits are identified:

$\varepsilon = 1$ **Elastic**

$\varepsilon = 0$ **Totally inelastic** .

The latter collision is 'putty-like' whereby the two bodies stick together. Since energy is no longer conserved, Equation (4.32) replaces Equation (4.23), which again must be combined with Equation (4.16) to obtain the final velocities v_1 and v_2. The result is

$$v_1 = \frac{m_1 u_1 + m_2\left[u_2 + \varepsilon\,(u_2 - u_1)\right]}{m_1 + m_2}\,, \tag{4.33}$$

$$v_2 = \frac{m_2 u_2 + m_1\left[u_1 + \varepsilon\,(u_1 - u_2)\right]}{m_1 + m_2}\,. \tag{4.34}$$

Simple generalisations then follow for the special cases examined in the previous section. For the stationary target ($u_2 = 0$) the recoil velocities are

$$v_1 = \left(\frac{m_1 - \varepsilon m_2}{m_1 + m_2}\right) u_1\,, \quad v_2 = \frac{m_1(1+\varepsilon)}{m_1 + m_2} u_1\,, \tag{4.35}$$

and for the equal approach collision $u_1 = -u_2 = u$ the recoil velocities are

$$v_1 = -\left[\frac{(2\varepsilon + 1)m_2 - m_1}{m_1 + m_2}\right] u\,, \quad v_2 = \left[\frac{(2\varepsilon + 1)m_1 - m_2}{m_1 + m_2}\right] u\,. \tag{4.36}$$

Equations (4.33) and (4.34) are the most general statements that can be made about one-dimensional two-body collisions, and allow us to calculate things like the kinetic energy transfer that occurs. However, in practice the calculations can become very unwieldy and it is more insightful, as well as algebraically simpler, to calculate quantities in a frame of reference which somehow respects an inherent symmetry in the collision geometry. We will study this next.

4.7 Centre-of-mass Frame

At the beginning of this chapter we noted that in two-body dynamics the motion naturally divides into the centre-of-mass motion and the 'internal' motion due respectively to the presence of external and internal forces. Even though for collisions the forces are just internal and only act for short times, the concept of dividing the motion is still useful. Although we are ultimately interested in one-dimensional collisions, as we saw in Section 4.4 even the process of establishing that the collision is one-dimensional means transforming from a frame in which the particles can be describing linear paths in up to three dimensions, to a frame in which the bodies collide and recoil along a line. We are now going to specialise that frame transformation to one where the origin of the 'one-dimensional' frame is the centre-of-mass of the system. In this way, the centre of mass motion will be eliminated and all that will remain will be a symmetric collision in which the momenta of the bodies are equal and opposite. Of course it will often be the case that we are interested in calculating quantities in the original frame. The point here is that it is often a better strategy to transform to the centre-of-mass frame, see what happens there and then transform back, than to work directly in the original frame. Since we are transforming from a general inertial frame, it will be necessary to use full vector notation. So, in all generality:

- The **centre-of-mass frame** is an inertial frame in which the position vector of a particle, $\mathbf{r}^*$, is measured relative to the centre of mass

$$\mathbf{r}^* = \mathbf{r}^{\text{lab}} - \mathbf{R}\,. \tag{4.37}$$

Here $\mathbf{r}^{\text{lab}}$ is the position vector of the particle in the original frame, and $\mathbf{R}$ is the location of the centre of mass as given by Equation (4.4), with $\mathbf{r}_1$ in that formula now being replaced by $\mathbf{r}_1^{\text{lab}}$ etc. The position vector of particle 1, for example, is $\mathbf{r}_1^* = \mathbf{r}_1^{\text{lab}} - \mathbf{R}$ (see Figure 4.6). The velocity of particle 1 relative to the centre of mass is given by

$$\dot{\mathbf{r}}_1^* = \dot{\mathbf{r}}_1^{\text{lab}} - \dot{\mathbf{R}}\,, \tag{4.38}$$

or

$$\mathbf{v}_1^* = \mathbf{v}_1^{\text{lab}} - \mathbf{V}\,, \tag{4.39}$$

where $\mathbf{V} = \dot{\mathbf{R}}$ is the velocity of the centre of mass.

Before we become too bogged down with all these superscripts, subscripts and dots – a notation students often find confusing, but it's not too bad really if you bear in mind the key

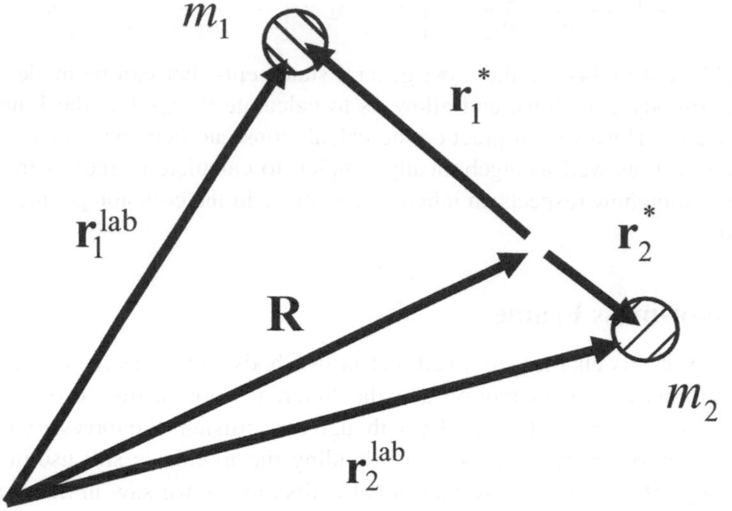

Figure 4.6 Connection between lab and centre-of-mass descriptions of the position of two bodies.

diagram Figure 4.6 – it is worthwhile considering a couple of examples that treat the centre-of-mass frame from a conceptual viewpoint. Firstly, we will revisit the example of the 'equal approach' collision of the last section, which we will solve neatly by stepping in and out of the centre-of-mass frame:

Example 1: Two unequal masses collide in the laboratory with equal and opposite velocities u and $-u$. Show that the lighter mass cannot recoil with a velocity greater than $3u$.

Since a frame has been chosen in which the bodies are already approaching along a line, transformation to the centre-of-mass frame is along one direction only and for the moment we can drop the vector notation. Then in the limit $m_1 \gg m_2$, say, we have that the centre-of-mass velocity $V \approx u$, since the centre of mass is more or less located at the larger mass. The velocities of the bodies before the collision in the centre-of-mass frame are, according to Equation (4.39):

$$u_1^* = u_1^{\text{lab}} - V \approx u - u = 0\,,$$

$$u_2^* = u_2^{\text{lab}} - V \approx -u - u = -2u\,. \tag{4.40}$$

Thus in the centre-of-mass frame we have reduced the problem to a 'brick wall rebound' of the lighter mass against the heavier mass (see Figure 4.7).

In this frame the recoil velocities are then

$$v_1^* = 0 \text{ and } v_2^* = 2u\,. \tag{4.41}$$

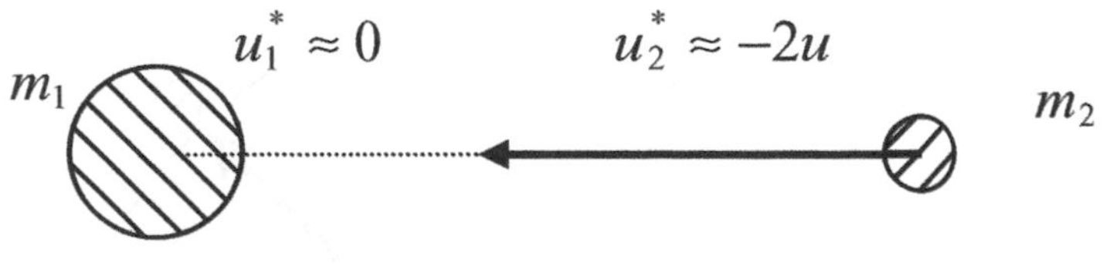

Figure 4.7 Equal approach in the centre-of-mass frame.

Using Equations (4.39) again to transform back into lab frame we obtain for the velocities after the collision

$$v_1^{\text{lab}} = v_1^* + V = 0 + u = u\,,$$
$$v_2^{\text{lab}} = v_2^* + V = 2u + u = 3u\,, \tag{4.42}$$

which is the result we sought.

Our second example, which actually concerns an oblique collision, explains why snooker balls always scatter at 90° and is a beautiful exploitation of centre-of-mass frame reasoning.

Example 2: A ball strikes elastically an identical stationary ball on a table. Calculate the angle between the ball's velocity vectors after the collision.

Going straight into the centre-of-mass frame, Figure 4.8 shows the balls approaching each other with equal and opposite velocities $\mathbf{u}^*$ and $-\mathbf{u}^*$. Since both momentum and energy are conserved, the speeds of the masses are unchanged by the collision, but will now be following new equal and opposite velocities $\mathbf{v}^*$ and $-\mathbf{v}^*$. All velocity vectors are therefore radii $|\mathbf{u}^*| = |\mathbf{v}^*|$ of a circle as shown. Now step back into the lab frame and obtain Figure 4.9. The velocity vectors in the lab frame are found by adding $\mathbf{u}^*$, the velocity of the centre of mass as viewed from the lab, to the velocities in the centre-of-mass frame. The horizontal diameter now represents the incident velocity of the ball from the left (i.e. $\mathbf{u}_1^{\text{lab}} = 2\mathbf{u}^*$), whilst adding $\mathbf{u}^*$ to $\mathbf{v}^*$ yields $\mathbf{v}_1^{\text{lab}} = \mathbf{u}^* + \mathbf{v}^*$, the velocity of the incident ball after the collision. Finally, adding $\mathbf{u}^*$ to $-\mathbf{v}^*$ yields $\mathbf{v}_2^{\text{lab}} = \mathbf{u}^* - \mathbf{v}^*$, the velocity of the stationary ball after the collision, and this arrow has been placed in the upper half of the diagram as shown. Now it is clear that since $\mathbf{v}_1^{\text{lab}}$ and $\mathbf{v}_2^{\text{lab}}$ subtend a diameter they must be at right angles to each other. The balls always scatter at 90° to each other! Notice how we recover known results in limiting cases. As point P approaches A the incident ball is brought to rest and transfers all its energy and momentum to the second ball. As point P approaches B the incident ball just glances the stationary ball, hardly moving it (perpendicularly!) at all.

Now we proceed to analyse in the centre-of-mass frame generally. A useful coordinate to introduce is the *relative* coordinate between the two particles given by

$$\mathbf{r} \equiv \mathbf{r}_2^{\text{lab}} - \mathbf{r}_1^{\text{lab}} = \mathbf{r}_2^* - \mathbf{r}_1^*\,, \tag{4.43}$$

which, you will notice, is the same in both frames as any difference in origin is subtracted out. Solving Equation (4.43) and Equation (4.4) for $\mathbf{r}_1^{\text{lab}}$ and $\mathbf{r}_2^{\text{lab}}$ we obtain the lab positions in terms of the centre of mass and the relative displacement as

$$\mathbf{r}_1^{\text{lab}} = \mathbf{R} - \frac{m_2}{M}\mathbf{r}\,, \tag{4.44}$$

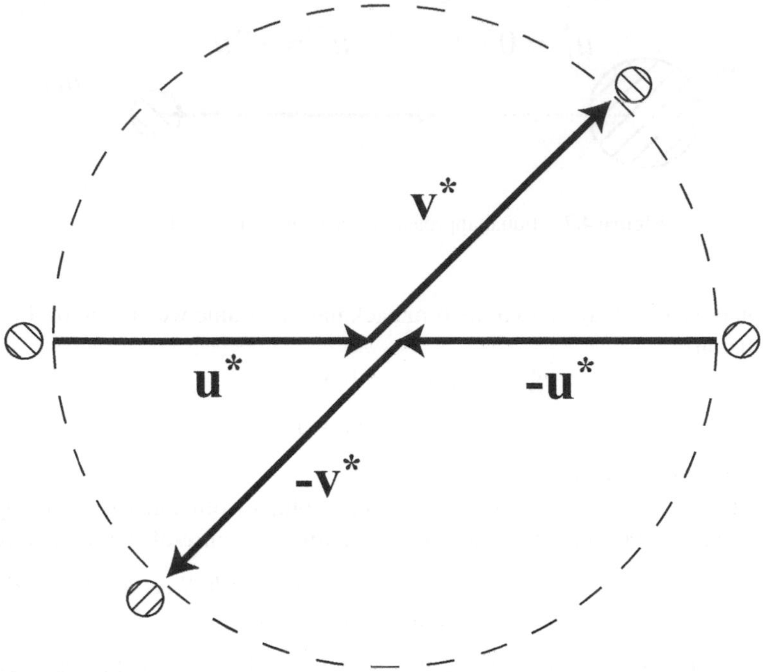

Figure 4.8 Two identical masses scattering as viewed from the centre-of-mass frame.

and

$$\mathbf{r}_2^{\rm lab} = \mathbf{R} + \frac{m_1}{M}\mathbf{r}\,. \tag{4.45}$$

We can now calculate the momenta of the particles in the centre-of-mass frame:

$$\mathbf{p}_1^* = m_1\dot{\mathbf{r}}_1^* = m_1\left(\dot{\mathbf{r}}_1^{\rm lab} - \dot{\mathbf{R}}\right)$$

$$= m_1\left(\dot{\mathbf{R}} - \frac{m_2}{M}\dot{\mathbf{r}} - \dot{\mathbf{R}}\right) = -\mu\dot{\mathbf{r}}\,, \tag{4.46}$$

a similar calculation yielding for the other particle $\mathbf{p}_2^* = +\mu\dot{\mathbf{r}}$. Thus the momenta are equal and opposite and of magnitude given by the reduced mass times the relative speed. Since the total momentum is always zero, the centre-of-mass frame is sometimes called the **centre-of-momentum** frame. The kinetic energies in the centre of mass frame are given by

$$K_1^* = \frac{|\mathbf{p}_1^*|^2}{2m_1} = \frac{\mu^2\dot{r}^2}{2m_1}\,, \quad K_2^* = \frac{|\mathbf{p}_2^*|^2}{2m_2} = \frac{\mu^2\dot{r}^2}{2m_2}\,, \tag{4.47}$$

where $\dot{r} = |\dot{\mathbf{r}}|$. By themselves these formulae are not very useful, but the total kinetic energy in the centre-of-mass frame, K^*, given by summing Equations (4.47) is worth remembering:

$$K^* = K_1^* + K_2^* = \frac{1}{2}\mu\dot{r}^2 = \frac{(p^*)^2}{2\mu}\,. \tag{4.48}$$

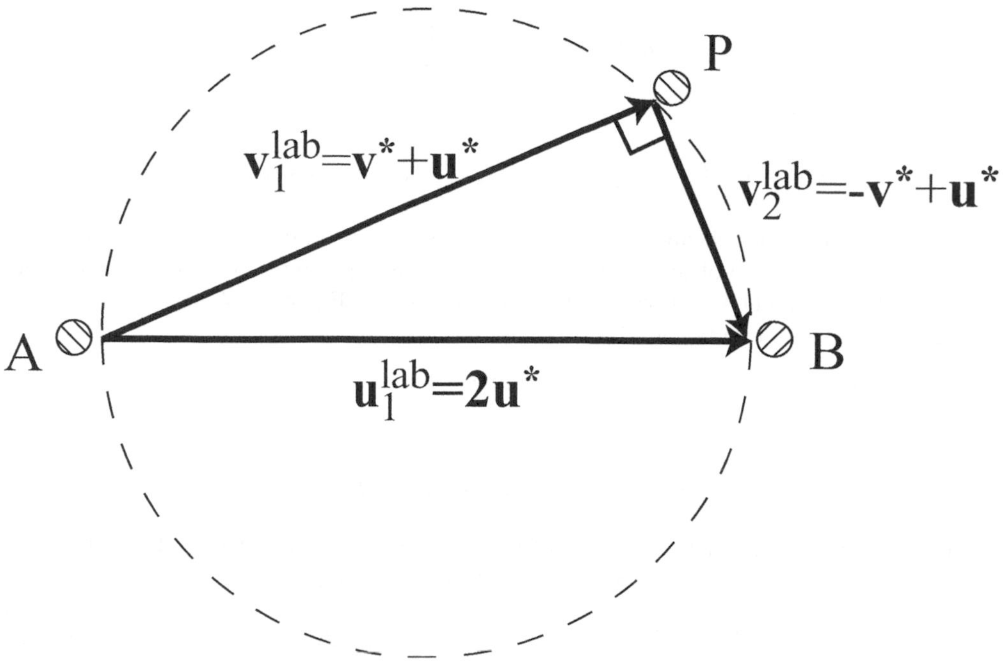

Figure 4.9 Two identical masses scattering as viewed from the lab frame.

All that remains is to relate the total lab kinetic energy, $K^{\rm lab}$, to the total centre-of-mass kinetic energy:

$$\begin{aligned} K^{\rm lab} &= \tfrac{1}{2}m_1\left|\dot{\mathbf{r}}_1^{\rm lab}\right|^2 + \tfrac{1}{2}m_2\left|\dot{\mathbf{r}}_2^{\rm lab}\right|^2 \\ &= \tfrac{1}{2}m_1\left|\dot{\mathbf{R}} - \tfrac{m_2}{M}\dot{\mathbf{r}}\right|^2 + \tfrac{1}{2}m_2\left|\dot{\mathbf{R}} + \tfrac{m_1}{M}\dot{\mathbf{r}}\right|^2 \\ &= \tfrac{1}{2}M\dot{R}^2 + \tfrac{1}{2}\mu\dot{r}^2, \end{aligned}$$

where Equations (4.44) and (4.45) have been used as well as $|\mathbf{a}|^2 = \mathbf{a}\cdot\mathbf{a} \equiv a^2$. We therefore have the important relationship

$$K^{\rm lab} = K^{\rm CM} + K^* \,, \tag{4.49}$$

where $K^{\rm CM}$ is the kinetic energy of the centre of mass.

Where is all this leading? Let me try to convince you that there really is some computational utility in using the centre-of-mass frame by way of an example:

Example: Calculate the change in kinetic energy for an inelastic 1-D collision.

Method 1: Working entirely in the lab frame:

The total kinetic energies before and after the collision are given respectively by

$$K^{\rm lab}_{\rm before} = \frac{1}{2}m_1u_1^2 + \frac{1}{2}m_2u_2^2 \,,$$

and

$$K^{\text{lab}}_{\text{after}} = \frac{1}{2}m_1v_1^2 + \frac{1}{2}m_2v_2^2 ,$$

so that the change is given by

$$\Delta K = K^{\text{lab}}_{\text{before}} - K^{\text{lab}}_{\text{after}} = \frac{1}{2}m_1(u_1^2 - v_1^2) + \frac{1}{2}m_2(u_2^2 - v_2^2) . \tag{4.50}$$

The unknown velocities, v_1 and v_2, are calculated using momentum conservation, Equation (4.16), and the definition of the coefficient of restitution, Equation (4.32), as before to yield Equations (4.33) and (4.34). These are then substituted into Equation (4.50). This is clearly an algebraically tedious procedure!

Method 2: Working via the centre-of-mass frame:

From Equation (4.49)

$$K^{\text{lab}}_{\text{before}} = K^{CM} + K^*_{\text{before}} , \quad K^{\text{lab}}_{\text{after}} = K^{CM} + K^*_{\text{after}} .$$

Since no external forces act, the centre-of-mass velocity, $\dot{R}$, is constant and therefore the centre-of-mass kinetic energy is unchanged by the collision. Hence

$$\Delta K = K^{\text{lab}}_{\text{before}} - K^{\text{lab}}_{\text{after}} = K^*_{\text{before}} - K^*_{\text{after}} , \tag{4.51}$$

Now $K^* = \mu\dot{r}^2/2$ (Equation (4.48)), and since the result of the collision is to reduce the relative approach speed, $\dot{r}$, by a factor of ε, then K^* must be reduced by a factor of ε^2:

$$K^2_{\text{after}} = \varepsilon^2 K^*_{\text{before}} . \tag{4.52}$$

Combining this with Equation (4.51) yields

$$\Delta K = (1 - e^2)K^*_{\text{before}} = (1 - \varepsilon^2)\frac{1}{2}\mu\dot{r}^2 ,$$

or

$$\Delta K = \frac{m_1m_2(1 - \varepsilon^2)(u_1 - u_2)^2}{2(m_1 + m_2)} , \tag{4.53}$$

and a lot of algebra has been finessed. Note that $\Delta K = 0$ if $\varepsilon = 1$, so that there is no change of kinetic energy for an elastic collision as expected.

The problems at the end of the chapter give more practice in using the centre-of-mass frame in calculations.

4.8 Rocket Motion

Our final example of two-body dynamics is rather different. Imagine sitting on a trolley holding a rifle and firing it horizontally. The resultant two-body problem consists of taking the person–rifle–trolley combination as one body, and the bullet as the other. The 'before' state is the complete system just sitting there with zero momentum. Now as a result of an explosion in the rifle, the person–rifle–trolley part of the system exerts a force on the bullet, the bullet in its turn exerting an equal and opposite force on the rifle. These internal forces push the bullet and the trolley apart so that the 'after' state consists of the bullet travelling in one direction and the rest of the system moving oppositely. The small bullet travels very rapidly, whilst the trolley, being more massive, goes slowly, such that the overall momentum of the system remains zero. Now replace the rifle with a machine gun. When firing continuously, each bullet carries a little bit of momentum in one direction, whilst the person plus gun plus trolley receives an equal and opposite increment of momentum in the other direction. Although each shot only results in a small velocity increment of the trolley, when the gun has finished firing, all the velocity increments have added up to give it considerable velocity. Now since the bullet mass is always much less than the mass of the rest of the system, surely there must be a continuum limit of this process which describes a continuous *flow* of momentum away from a system which itself continuously increases its velocity (i.e. accelerates) in opposition. This is basically how a rocket works, accelerating in response to the burning of chemical fuel. Note that the fuel may actually comprise a significant fraction of the system mass, so that when all the fuel is burnt, the remaining mass will be substantially less. Somehow this mass variability must be taken into account.

Consider Figure 4.10, which depicts a rocket in free space viewed from an inertial frame such that at time t it is travelling with velocity v. Let the rocket eject the fuel continuously with velocity $(-v_{\rm ex})$ *relative to the rocket*, so that relative to the inertial frame the fuel velocity is $v - v_{\rm ex}$. In time dt, let the mass of the rocket change from m to $m + dm$ so that the mass of the fuel ejected is $-dm$. In this time the velocity changes from v to $v + dv$ (see Figure 4.10(b)). Note that in this formulation dm is negative (the reason for choosing the sign in this way will soon become apparent). Comparing Figures 4.10(a) and 4.10(b),

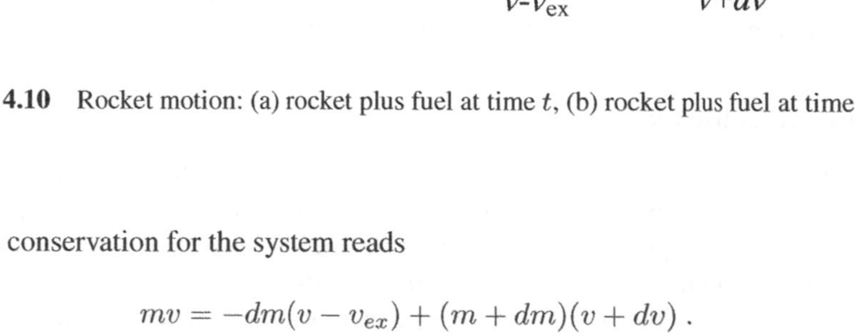

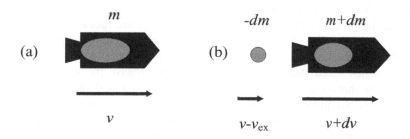

Figure 4.10 Rocket motion: (a) rocket plus fuel at time t, (b) rocket plus fuel at time t+dt.

momentum conservation for the system reads

$$mv = -dm(v - v_{ex}) + (m + dm)(v + dv)\,. \qquad (4.54)$$

In expanding the right-hand side of this expression we can ignore the term $dmdv$ as this is second order in small quantities, so that cancelling and rearranging we obtain

$$mdv = -v_{\rm ex}dm. \qquad (4.55)$$

Dividing by dt we see that the left-hand side is a mass times an acceleration, so that the right-hand side must represent a force, and is called the **thrust** experienced by the rocket:

$$m\frac{dv}{dt} = -v_{\text{ex}}\frac{dm}{dt} \, . \tag{4.56}$$

Remember that dm is negative[1] so that the thrust is positive. Note that since no external force is acting this is equivalent to an internal force acting within the system. Equation (4.55) can be used to determine how the velocity of the rocket increases as its mass changes:

$$\frac{dm}{m} = -\frac{dv}{v_{\text{ex}}} \; ; \quad \int_{m_i}^{m_f} \frac{dm}{m} = -\int_{v_i}^{v_f} \frac{dv}{v_{\text{ex}}} \, ,$$

$$[\ln(m)]_{m_i}^{m_f} = -\left[\frac{v}{v_{\text{ex}}}\right]_{v_i}^{v_f} \; ; \quad \ln\left(\frac{m_f}{m_i}\right) = \frac{v_i - v_f}{v_{\text{ex}}} \, ,$$

$$v_f - v_i = v_{\text{ex}} \ln\left(\frac{m_i}{m_f}\right) \, , \tag{4.57}$$

$$m_i = m_f \exp\left(\frac{v_f - v_i}{v_{\text{ex}}}\right) \, . \tag{4.58}$$

Here the i and f subscripts on m and v refer to their initial and final values. We have figured out what the initial mass must be to achieve a given final rocket velocity. On account of the exponential relationship, to achieve a significant increase in the rocket's speed, the initial mass (including the fuel) must be *very much greater* than the rocket mass on its own. The speed boost $(v_f - v_i)$ depends on the exhaust velocity v_{ex} but is independent of the constant burn rate dm/dt. For chemical rockets $v_{\text{ex}} \leq 4\,\text{km}\,\text{s}^{-1}$.

Of course the process can be reversed so that sending the fuel out ahead of the rocket *decelerates* the rocket. The same analysis would then apply except that the exhaust velocity relative to the lab would now become $v + v_{\text{ex}}$ and finally we would obtain $v_f < v_i$.

It is sometimes stated that the rocket accelerates *because* the mass is changing, the argument saying that since momentum is constant, $d(mv)/dt = 0$ implies that $dv/dt \neq 0$ if $dm/dt \neq 0$. However, just changing the mass of a system per se does not lead to acceleration. The change in mass must be associated with some momentum being carried away, from which the remaining part of the system must then recoil.

4.9 Launch Vehicles

In the rocket analysis of the last section, gravity was ignored. However, gravity is clearly crucial for a rocket to escape from the Earth and so we will analyse it here (see Figure 4.11). The analysis leading to Equation (4.55) can now be interpreted as being a means to calculate the *change* in momentum in time dt as $(m + dm)(v + dv) + (-dm)(v - v_{\text{ex}}) - mv$, so that dividing this by dt gives us the rate of change of momentum, equal to the external force acting, $-mg$:

$$m\frac{dv}{dt} + v_{\text{ex}}\frac{dm}{dt} = -mg \, . \tag{4.59}$$

[1] If we had chosen dm to be positive, then we would be calculating the mass of fuel ejected rather than the change in mass of the rocket.

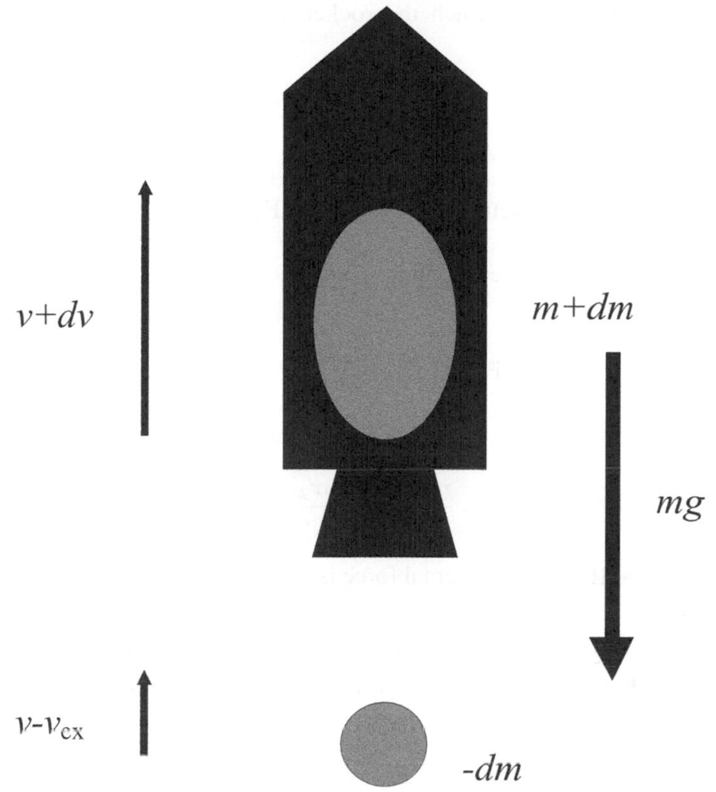

Figure 4.11 Rocket motion in the presence of gravity. This figure depicts rocket plus fuel at time *t+dt*.

Evidently to achieve a positive acceleration, $dv/dt > 0$, requires $v_{\text{ex}}|dm/dt| > mg$, or the thrust must overcome the rocket's weight. Unlike the previous case, the burn rate is now significant and, in order to take off, the launch vehicle must burn its fuel *quickly*. Launch is so spectacular precisely for this reason. Dividing by m and integrating from the initial state at $t = 0$ to the final time t

$$\int_{v_i}^{v_f} dv = -v_{\text{ex}} \int_{m_i}^{m_f} \frac{dm}{m} - g \int_0^t dt \,, \tag{4.60}$$

$$v_f - v_i = v_{\text{ex}} \ln\left(\frac{m_i}{m_f}\right) - gt \,. \tag{4.61}$$

The velocity boost is therefore similar to Equation (4.57) except that now it is diminished by the effect of gravity. We have assumed that gravity is constant during the burn so that the distance travelled is much less than the radius of the Earth. Actually this approximation is not too bad since the burn required to overcome gravity and reach escape velocity is necessarily rapid (see comments above), after which the rocket travels ballistically, the analysis of Section 2.6 then applying.

4.10 Summary

Section 4.2. For two masses, m_1 and m_2, the external forces acting on each body ($\mathbf{F}_1^{\text{Ext}}$ and $\mathbf{F}_2^{\text{Ext}}$) and the internal forces acting between them ($\mathbf{F}_{12}$ and $\mathbf{F}_{21}$) combine to yield the equations of motion,

$$m_1\ddot{\mathbf{r}}_1 = \left(\mathbf{F}_{12} + \mathbf{F}_1^{\text{Ext}}\right) , \tag{4.1}$$

and

$$m_2\ddot{\mathbf{r}}_2 = \left(\mathbf{F}_{21} + \mathbf{F}_2^{\text{Ext}}\right) . \tag{4.2}$$

Adding these yields $(m_1 + m_2)\ddot{\mathbf{R}} = \mathbf{F}_1^{\text{Ext}} + \mathbf{F}_2^{\text{Ext}}$ (Equation (4.5)), where

$$\mathbf{R} = \frac{m_1\mathbf{r}_1 + m_2\mathbf{r}_2}{m_1 + m_2} , \tag{4.4}$$

defines the centre of mass. If the total external force is zero, then the momentum of the centre of mass is constant.

Section 4.3. When the external force on each mass is the same, subtracting the equations of motion yields $\mu\ddot{\mathbf{r}} = \mathbf{F}^{\text{int}}$, Equation (4.10), where $\mathbf{r} \equiv \mathbf{r}_2 - \mathbf{r}_1$, and the reduced mass is defined through

$$\mu \equiv \left(\frac{1}{m_1} + \frac{1}{m_2}\right)^{-1} . \tag{4.9}$$

Section 4.4. Collisions occur when two bodies interact for a short time. A one-dimensional collision is where the two bodies approach and scatter along the same direction. Momentum conservation is expressed as:

$$m_1u_1 + m_2u_2 = m_1v_1 + m_2v_2 . \tag{4.16}$$

Section 4.5. Elastic collisions are characterised by energy conservation

$$K_{\text{initial}} = \frac{1}{2}m_1u_1^2 + \frac{1}{2}m_2u_2^2 = \frac{1}{2}m_1v_1^2 + \frac{1}{2}m_2v_2^2 = K_{\text{final}} . \tag{4.19}$$

Section 4.6. Inelastic collisions are characterised by the coefficient of restitution, ε:

$$v_2 - v_1 = \varepsilon\left(u_1 - u_2\right) . \tag{4.32}$$

Equation (4.16) can be combined with either Equation (4.19) or (4.32) to solve for the final velocities v_1 and v_2 (Equations (4.24) and (4.25), or (4.33) and (4.34)).

Section 4.7. The centre-of-mass frame is one in which positions are measured relative to the centre of mass (Figure 4.6):

$$\mathbf{r}^* = \mathbf{r}^{\text{lab}} - \mathbf{R} . \tag{4.37}$$

Calculations of kinetic energy transfer in collisions are often best performed in the centre-of-mass frame.

Section 4.8. The dynamics of a rocket ejecting fuel at speed v_{ex}, in the absence of gravity, is expressed through the equation

$$m\frac{dv}{dt} = -v_{\text{ex}}\frac{dm}{dt} , \tag{4.56}$$

the solution of which is

$$m_i = m_f \exp\left(\frac{v_f - v_i}{v_{\text{ex}}}\right) . \tag{4.58}$$

Section 4.9. A launch vehicle extends this analysis to include gravity, resulting in the velocity boost

$$v_f - v_i = v_{\text{ex}} \ln\left(\frac{m_i}{m_f}\right) - gt . \tag{4.61}$$

4.11 Problems

4.1 Two masses, $m_1 = 1\,\text{kg}$, $m_2 = 3\,\text{kg}$ are located in the (x, y)-plane at positions $\mathbf{r}_1 = 1\mathbf{j}$ m and $\mathbf{r}_2 = -2\mathbf{i}$ m respectively ($\mathbf{i}$ and $\mathbf{j}$ are unit vectors in the x and y directions respectively). Calculate the reduced mass of m_1, m_2 and the (x, y)-coordinates of their centre of mass.

4.2 If m_1 and m_2 of Problem 4.1 have velocities $\mathbf{v}_1 = 2\mathbf{i} + 3\mathbf{j}(\text{m s}^{-1})$ and $\mathbf{v}_2 = 1\mathbf{i}(\text{m s}^{-1})$ find: (a) the velocity of the centre of mass, (b) the velocity of m_1 relative to m_2, (c) the momenta of m_1 and m_2 in the centre-of-mass frame and (d) the total kinetic energy of particles 1 and 2 in the centre-of-mass frame.

4.3 A rubber ball is released from rest at time $t = 0$ from a height of 5 m. It bounces from the ground with a coefficient of restitution given by $\varepsilon = 0.5$. Calculate the time after which the bouncing ball finally comes to rest.

4.4 A man of mass 60 kg stands at one end of a 20 m plank (with a mass of 20 kg) which lies on a frozen pond. If he walks to the other end of the plank and the friction between plank and ice is negligible, how far does he move relative to the pond?

4.5 Two masses, m_1 and m_2, travelling at velocities u_1 and u_2 in the same straight line undergo a completely inelastic collision (coefficient of restitution $\varepsilon = 0$). Show that the loss of kinetic energy is equal to the kinetic energy of the reduced mass, μ, moving with the original relative velocity.

4.6 A mass m_1, of 2 kg moving with velocity $-3\mathbf{j}$ m s^{-1} and a mass m_2, of 4 kg moving with velocity $-2\mathbf{i}$ m s^{-1} collide at the origin in the laboratory frame.

(a) Draw a sketch which shows the positions and directions of the two masses and of the centre of mass, 1 s before the collision.

(b) Calculate the x and y components of the centre-of-mass velocity, $\dot{\mathbf{R}}$, and show that its magnitude is $5/3$ m s^{-1}.

(c) Calculate the x and y components, $(\dot{r}_1^*)_x$ and $(\dot{r}_1^*)_y$ of the velocity of mass m_1 in the centre-of-mass frame, before the collision and show that the speed in this frame is $\dot{r}_1^* = (52/9)^{1/2}$ m s^{-1}.

(d) Show similarly that the speed of m_2 in the centre-of-mass frame is $\dot{r}_2^* = (13/9)^{1/2}\ \mathrm{m\,s^{-1}}$.

(e) Show that the total kinetic energy, K^*, in the centre-of-mass frame before the collision, is $26/3\,\mathrm{J}$.

(f) Show that the total kinetic energy, K, in the laboratory frame before the collision is $17\,\mathrm{J}$.

(g) If the collision reduces the relative velocity by a factor of two, by what factor is the kinetic energy in the centre-of-mass frame reduced? Hence show that the final total kinetic energy measured in the laboratory frame of reference is $10.5\,\mathrm{J}$.

4.7 In a nuclear reactor the neutrons emitted in the fission of the uranium fuel are too fast to cause subsequent fission reactions. It is the task of the moderator to slow the neutrons down such that their probability of causing fission is increased sufficiently to sustain a chain reaction. In this problem the moderator contains deuterium. A neutron (n) which is produced with a speed of $10^4\,\mathrm{m\,s^{-1}}$ collides and scatters elastically from a stationary deuteron (d). The masses of a neutron, a deuteron and a carbon nucleus are respectively $m_n = 1.7 \times 10^{-27}\,\mathrm{kg}$; $m_d = 3.4 \times 10^{-27}\,\mathrm{kg}$; $m_C = 20.4 \times 10^{-27}\,\mathrm{kg}$. Calculate:

(a) the speed of the centre of mass (of the n–d system) in the laboratory frame.

(b) the speed before the collision of the neutron in the centre-of-mass frame.

(c) the speed before the collision of the deuteron in the centre-of-mass frame.

(d) the ratio of the total kinetic energy in the laboratory frame to that in the centre-of-mass frame.

(e) the maximum fractional kinetic energy which can be transferred from the neutron to the deuteron (this occurs when the particles collide one-dimensionally).

(f) the number of collisions required to reduce the speed of the neutron to less than $10^3\ \mathrm{m\ s^{-1}}$ if the average fraction of kinetic energy transferred is one half the maximum value due to the particles colliding obliquely.

(g) the number of collisions required if a graphite (carbon) moderator had been used instead.

4.8 A rocket expels burnt fuel at a constant speed of $3000\,\mathrm{m\,s^{-1}}$ relative to the rocket. Calculate the mass of fuel that must be burnt to boost the speed of a 1 kg mass from rest up to 1% of the speed of light. Neglect any forces due to gravity ($c = 3 \times 10^8\,\mathrm{m\,s^{-1}}$).

4.9 A rocket accelerates in free space by ejecting fuel at a constant relative speed, v_{ex}. It consists of two stages and three-quarters of its total mass is fuel. It accelerates from rest until two-thirds of its fuel is burnt. The first stage is then detached. Fuel accounts for 80% of the mass of the second stage. Show that the final velocity of the second stage will be $v = v_{\mathrm{ex}} \ln 10$. Show that a single stage rocket of the same mass and burning the same total mass of fuel would attain a final velocity of only $0.6v$.

4.10 A rocket travels in free space at a speed of $1\,\mathrm{km\,s^{-1}}$. It carries a payload of 100 kg and the best available chemical propellant, which can eject burnt fuel at a speed of $3\ \mathrm{km\ s^{-1}}$. The rocket engineers (whose advice the management always heed!) demand, for safety, a fuel-containing structure with a mass at least 10% of the initial fuel mass.

(a) Estimate the minimum mass of fuel required to boost the speed of the payload to: (i) 4 km s^{-1}, (ii) 7 km s^{-1}, (iii) 10 km s^{-1}.

(b) The payload must, however, reach a speed of 13 km s^{-1}. To achieve this, a two-stage rocket is built. Each stage satisfies the engineering safety requirements. If the speed boost provided by each stage is equal, estimate the mass of the fuel required for each stage.

(c) This two-stage rocket, with fuel loads as in part (a), is now to be used as a launch vehicle. Write down the equation governing vertical rocket motion in the Earth's gravitational field and hence determine the minimum rate at which the fuel must be burnt (kg s^{-1}) to ensure take-off from a launch pad on Earth (assume $g = 10$ m s^{-2} independent of height).

(d) If the burn rate is constant, estimate the time at which the first stage should be detached.

(e) Estimate the speed of the payload after the second stage burn is complete if the burn rate is the same as used in the first stage.

5

Relativity 1: Space and Time

5.1 Why Relativity?

Newton's theory was, and is, outstandingly successful, so why bother to refine it? First, as physicists we seek the best possible description even if in practice we use an approximate theory. However, more importantly it turns out that in situations where speeds approach that of light ($v \sim c$) the laws of physics are *very* different. For example, if a body moves to the right with velocity v' in a frame moving to the right with a velocity u (which we will refer to as the rocket frame), then, according to relativity, the laboratory observer will see the body move with velocity, v, given by

$$v = \frac{v' + u}{1 + uv'/c^2} . \tag{5.1}$$

This formula will be proved later in this chapter. For now let us note that it yields the familiar Newtonian velocity addition law $v = u + v'$ if $uv' \ll c^2$ (the correspondence principle) and yields unfamiliar results if *either* the particle velocity, v', *or* the frame velocity, u, approach the velocity of light, c.

The special theory of relativity to be developed in this chapter is the best theory we have for dynamics in inertial frames. The mathematics of special relativity is simple – no more than elementary calculus is necessary. General relativity, concerning the relativistic description of gravity, is much more mathematically forbidding and will not be discussed here.

That there was something wrong with Newtonian physics began to emerge towards the end of the nineteenth century. The story of Bern's most famous patent office clerk has been told often enough to bear its omission here.[1] It is, however, a mark of extraordinary creative endeavour, largely by one man, that the refinement of Newtonian physics was achieved before it was possible to observe particle motion at speeds approaching the speed of light. As always, experiments guided the creative force to find the new theory, but the experiments in this case were unusual. In a sense they were trivial and, in the development of relativity, *conceptual* insight, through a fundamental reassessment of the underlying notions of space and time, played an unusually significant role. Questions like 'What do we really mean by an "observer"?', or 'How do given inertial observers measure separations in space and time?', or 'How are the measurements of one inertial observer related to those of another?' will be to the fore in this chapter. Careful consideration of these issues will lead directly to unfamiliar results like Equation (5.1), and cause us to refine, or *redefine*, momentum and energy.

[1]The definitive biography of Einstein is *Subtle is the Lord*, by Abraham Pais, Oxford University Press, 1982.

Classical Mechanics – Second Edition From Newton to Einstein: A Modern Introduction Martin W. McCall

5.2 Galilean Relativity

Before embarking on the study of Einstein's relativity, let us recall which relativistic principles operate within a Newtonian framework. According to Newton's first law, when no external forces act in an inertial frame, a body moves with constant velocity, $\mathbf{v}$, say. Now observe the same body from a frame that is moving with velocity $\mathbf{u}$ with respect to the first frame, so that the velocity of the body observed from the new frame is, according to Newton and Galileo, $\mathbf{v}' = \mathbf{v} - \mathbf{u}$. Newton's first law is still operative in the new frame, but the constant velocity entering into the statement of the law is different. Similarly, a ball struck towards the boundary of a cricket field follows a parabolic trajectory. However, the precise form of the trajectory depends on the frame of reference, so that a fielder in pursuit with the same horizontal velocity as the ball, for example, will perceive the motion of the ball to be simply 'up' and 'down' in the vertical direction. The key point is that the acceleration of the ball is $g = 9.8\,\mathrm{m\,s^{-2}}$ vertically downwards in *both* frames. Thus Newton's law in *both* frames is $\mathbf{F} = m\ddot{\mathbf{r}} = m\mathbf{g}$. What causes the trajectories to be different are the distinct initial velocities in each frame.

To sum up, Newton's laws take the same form in different inertial frames of reference – or I can hold a cup of coffee equally comfortably in my laboratory or in an aeroplane. Indeed if there were no turbulence and I couldn't see out of the window, I might believe that the aeroplane were still stationary so that another expression of the Galilean principle of relativity is

Mechanics experiments cannot distinguish between inertial frames.

Now try the coffee-cup experiment on a roundabout. The coffee sloshes about and it is immediately apparent that I am no longer in an inertial frame. Thus the Galilean relativity principle is strictly limited to the inability of mechanics experiments to distinguish between *inertial* frames (recall that an inertial frame is one in which Newton's first law holds). The actual value of, for example, a particle velocity, depends on the frame chosen, but the *form* of Newton's laws is the same in *all* inertial frames. The situation is completely analogous to confirming Pythagoras' theorem on a right-angled triangle with respect to two Cartesian coordinate systems that are rotated with respect to each other. The coordinates of the triangle's vertices are different in each coordinate frame, but this does not affect the validity of Pythagoras' theorem. Quantities which take different values in different frames of reference, but which combine to form a law which is applicable in all frames, are referred to as **covariant** quantities. The covariant quantities in Newton's laws are the position and velocity of particles. There are some quantities, however, which do stay the same from frame to frame. Mass, charge and the number of particles all have values which are unaffected by uniform relative motion, and are referred to as **invariant** quantities. In the Pythagoras example the distances between the vertices of the triangle are all invariant.

Try to hold on to the geometric 'analogy' as it really helps in understanding frame changes in relativity. I say 'analogy' in quotes, because it turns out not to be an analogy at all. Frame changes in relativity *are* geometric in character, provided the concept of geometry is generalised beyond the geometry learnt at school.

5.3 The Fundamental Postulates of Relativity

Now to what extent can the concept of the indistinguishability of inertial frames be extended to other areas of physics: perhaps a nonmechanical experiment can determine who is *really* at rest? Figure 5.1(a) shows a coil moving between the poles of a bar magnet. As the coil passes through, the charges within the wire experience the magnetic field as a result of which they feel a force along the wire and create a current pulse recorded on the ammeter. Figure 5.1(b) shows a different experiment where the magnet moves over a stationary coil. Again, a current pulse is observed, though the physical description is different. The classical description is that as the magnet reaches the coil, an e.m.f. is induced as a result of the magnetic flux linked through the coil increasing as the magnet passes over. Traditionally we think of the forces on moving charges in magnetic fields and electromagnetic induction as distinct physical phenomena. However, experiments show that the recorded current pulse is the same in both cases. Indeed one suspects that this should be so from the symmetry between Figs. 5.1(a) and 5.1(b). Thus this attempt to distinguish between inertial frames fails. It turns out that

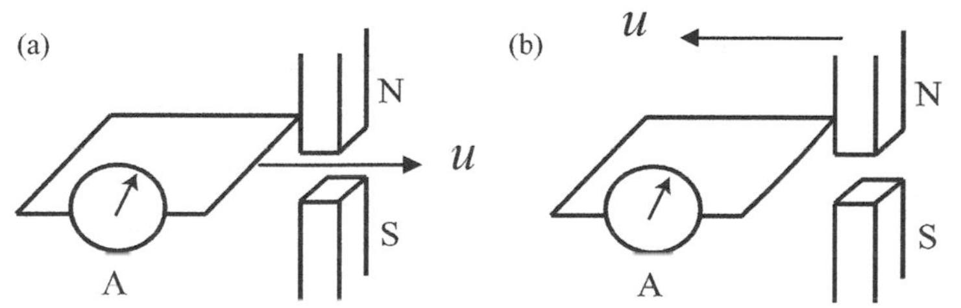

Figure 5.1 Relative motion between a coil and a bar magnet: (a) coil moving, magnet stationary; (b) coil stationary, magnet moving.

all physical phenomena are like this. Observe an experiment in a laboratory from a moving (= 'rocket') frame and the results are the same as when the experiment is performed in the rocket frame and viewed from a stationary (= 'laboratory') frame. No matter how disparate the physical description, whenever the experiment is observed from frames in relative motion, the results cannot be used to distinguish between inertial frames. Einstein, in 1905, elevated this to a fundamental physical principle applicable to all of Nature:

The laws of Nature are the same for all inertial observers.

We emphasise again the distinction between covariance and invariance inasmuch as it relates to the relativity principle. If the experiment is at rest in the laboratory and observed from a rocket, then we *expect* the value of certain quantities such as velocities to be different between frames. The relativity principle thus says two different but related things: first, when observing a *single* experiment from different frames we obtain distinct values for various quantities, but they must combine in such a way as to produce laws of physics *that are identical in form.* Secondly, if *two* identical experiments are at rest respectively in the lab and rocket frames, then the rocket observer recording the experiment at rest in the lab must obtain the same results as the lab observer records for the experiment at rest in the rocket.

Consider another example, this time from optics. Imagine that light propagates much like sound does through air, so that in measuring the time taken to travel between source and observer (presumed stationary with respect to each other), the speed of the medium supporting its propagation must be taken into account. If the wind is blowing towards the source of the sound then it will take longer for the sound to reach the receiver than when the wind blows in the opposite direction. We do not know at this stage what the medium might be that supports light propagation, but provisionally we shall call it the **ether**. The ether is at rest in Newton's frame of absolute space, so that as the Earth orbits the Sun it is travelling at $u = 3 \times 10^4\,\mathrm{m\,s^{-1}}$ through the ether. Now if the notion of having to take account of the ether wind is correct, the light should travel at different speeds in different directions in an Earthbound laboratory, according to whether it is travelling parallel or perpendicular to the ether wind. This is the basis of the famous Michelson–Morley experiment (1887), the principles of which are illustrated in Figure 5.2. The first figure shows the 'cross-stream' case

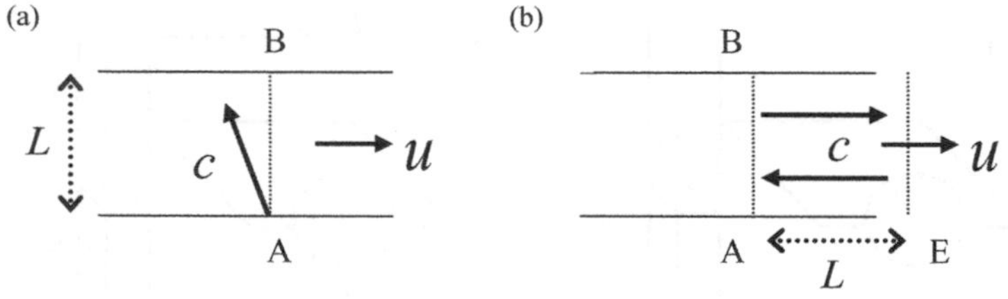

Figure 5.2 The basis of the Michelson–Morley experiment: (a) light travelling cross-stream to the ether; (b) light travelling along-stream with the ether.

where the light is directed such that its resultant velocity is perpendicular to the direction of the ether wind. If we assume standard velocity addition then the magnitude of this resultant velocity is $(c^2 - u^2)^{1/2}$ and the time taken for the round trip A–B–A ($= 2L$) is given by

$$t_{\mathrm{perp}} = \frac{2L}{\sqrt{c^2 - u^2}} \,. \tag{5.2}$$

For the 'along-stream' case (Figure 5.2(b)), on the other hand, the speed for the outbound journey for a $2L$ round trip is $c + u$, and for the return leg it is $c - u$, so that the overall journey time for the round trip A–E–A is given by:

$$t_{||} = \frac{L}{c+u} + \frac{L}{c-u} = \frac{2Lc}{c^2 - u^2} \,. \tag{5.3}$$

Taking the ratio

$$\frac{t_{||}}{t_{\mathrm{perp}}} = \frac{c}{\sqrt{c^2 - u^2}} = \frac{1}{\sqrt{1 - u^2/c^2}} \equiv \gamma \tag{5.4}$$

and plainly $t_{||} > t_{\mathrm{perp}}$, that is the along-stream journey takes longer than the cross-stream journey. The factor by which $t_{||}$ exceeds t_{perp} is the dimensionless number γ which occurs frequently in relativity. It depends only on the ratio u/c, the speed of the Earth through the ether, measured in units of the speed of light.

The Michelson–Morley experiment was based on an effective distance $L = 22$ m, so that putting in the numbers yields $t_{||} = 146.6666681$ ns and $t_{\rm perp} = 146.6666674$ ns or a time difference of $\Delta t = 7.3 \times 10^{-16}$ s which can be expressed fractionally as $(t_{||} - t_{\rm perp})/t_{\rm perp} = \Delta t/t = 10^{-8}$. The experiment was configured so that this extremely small time difference was converted to a path difference that could be detected by interfering the light travelling along the different paths. For light of wavelength λ, an interferometer can detect a path difference of approximately $\lambda/20$. This corresponds to a fractional path difference of

$$\frac{c\Delta t}{ct} = \frac{\lambda/20}{L}\,, \tag{5.5}$$

which yields, at the wavelength $\lambda = 589$ nm used

$$\frac{\Delta t}{t} = \frac{589 \times 10^{-9}}{(20 \times 22)} = 1.3 \times 10^{-9}\,. \tag{5.6}$$

Thus the small time difference should be detectable. In fact Michelson and Morley detected no time difference and the speed of the ether wind is zero. Now it might be that the ether is locally 'dragged along' by the Earth, and the null result of the Michelson–Morley experiment is because of this. This would, however, imply aberrational effects for light approaching the Earth from the distant stars that are not observed. Moreover, when the experiment was performed at different times of the year, when the ether wind would be blowing in different directions, still the result was the same. Whatever way we look at it, the null result of the Michelson–Morley experiment confirms that the velocity of light is **isotropic**, or the same in all directions. This most famous null experiment strongly suggests that the absolute space within which the ether is supposed to reside, and against which all inertial frames might be 'marked', is actually a fiction. There is no ether! The fact that the same null result is obtained at different times of the year (when the experiment is effectively being performed in different inertial frames) is consistent with the principle of relativity. If this were not the case then the frames would be distinguished. Notice that the experiment says nothing about the actual *value* of the speed of light in different inertial frames. It may be that in the inertial frame with which the Earth is comoving in January, the speed of light is $2.5 \times 10^8\,{\rm m\,s^{-1}}$. For the frame in June it might be $3 \times 10^8\,{\rm m\,s^{-1}}$. Both would produce a null result in the Michelson–Morley experiment provided the speed of light is isotropic. It requires *another experiment* to confirm that in fact the speed of light is the same in all inertial frames. This was provided later by the Kennedy–Thorndike experiment (1932) which showed, by direct measurements of the speed of light, that it is $c = 2.99792458 \times 10^8\,{\rm m\,s^{-1}}$, independent of the frame of observation.

In fact the theory of electromagnetism predicts, or at least strongly suggests, that light is a form of radiation due to a mutual coupling between electric and magnetic fields. The theory shows that in vacuum the speed of the radiation is given by $c = (\varepsilon_0\mu_0)^{-1/2}$ where ε_0 and μ_0 are respectively the permittivity and permeability of free space. Provided we accept that light is electromagnetic radiation then its speed *must* be the same for all inertial observers by the principle of relativity. If it were not so then the fundamental constants of nature (in this case ε_0 and μ_0) would change as we observed the speed of light from different inertial frames, and the frames would then be distinguished. The fundamental constants of nature, including c, are invariant quantities.

Already this leads to unfamiliar results. Consider a laser pulse fired along a train moving at speed u as in Figure 5.3. What speed does an observer on the platform measure? From what we have just seen, he, like his colleague on the train, must measure the value c, independent of the speed of the train u. If we believed Newtonian theory we would be forced to write

Figure 5.3 A light pulse travelling along a moving train.

something like $c = c + u$! Clearly there is something pretty fundamental at work here. We can see how it actually works by referring to the velocity addition law quoted at the beginning of this chapter (Equation (5.1)). Setting $v' = c$ yields $v = c$ independent of u. The speed of other bodies on the train will look different from different frames, but that is all right since they are not like light which is a special fundamental speed provided by Nature.

5.4 Inertial Observers in Relativity

In the light (excuse the pun!) of all this, we must reassess some of the basic notions about inertial frames introduced in Chapter 1 as we must be certain that the frames are constructed in a way that is consistent with the principle of relativity. In particular, we need to demonstrate that a given free test particle moves with constant velocity, to confirm that a given frame is inertial. The frame must be furnished with a coordinate system that might be represented as a latticework of metre sticks to measure the particle's various locations as it moves. However, we must also measure time. We already said in Chapter 1 that our clock must be 'good', but one good clock is not enough. What we really need is a *set of clocks* that can measure when a particle is *at specific locations* as it moves through the lattice. If velocity is to have any objective meaning then each metre rod must measure the same spatial extent, and each clock located at the vertices of the latticework must tick uniformly. However, a more pressing issue is how, practically, can we be sure that all the clocks in our putative inertial frame are synchronised? Starting them all together at the origin and then taking them to their various locations will not do as in moving them we may change the rate at which they tick. The answer is to place all the clocks in the frame first and then gauge their starting using the only standard measure of which we can be certain, the speed of light (see Figure 5.4). Place a clock at location (x, y, z) in the coordinate frame (Figure 5.4(a)) and set it to time $(x^2 + y^2 + z^2)^{1/2}/c$ seconds after 12 o'clock. Repeat this process for clocks placed at all the lattice vertices. Set the clock at the origin to 12 o'clock and set it running when a light pulse is emitted from the origin. As the spherical wave expands it sets off the clocks through

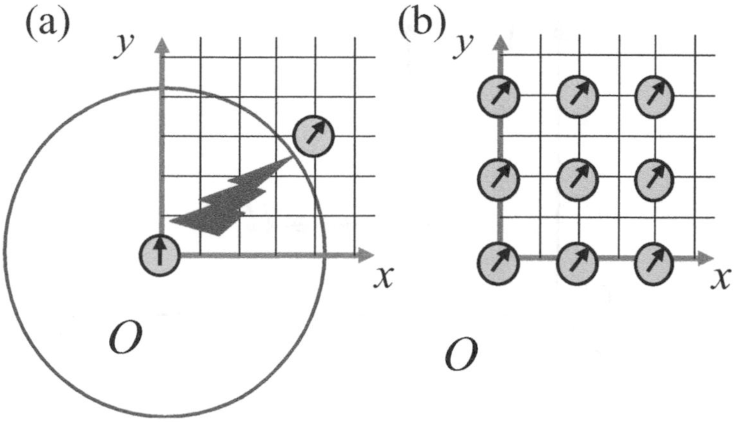

Figure 5.4 Synchronising clocks in an inertial frame.

which it passes. When the process is complete all the clocks in the frame are guaranteed to be synchronised as in Figure 5.4(b).

It is very important to appreciate that this latticework of rods and clocks represents a refinement of what we mean by an observer in an inertial frame. Each clock, labelled with its frame coordinates, records when particular **events** occur at its specific spatial location. The process of examining the trajectory of a particle is then *retrospective*, in the sense that a technician (or an undergraduate student!) tours the frame later, collecting the data punched out by the clocks for subsequent analysis. When we say **observer** in relativity we are really using shorthand for this retroactive process. If we were to think of an observer in the conventional sense, that is as an individual recording observations at a particular location, then we would need to take account of the time for signals to travel from an event to the observer. Our refined notion of observer obviates these complications.

One final comment before moving on. A constant velocity in an inertial frame implies that the test particles move in straight lines. But what is 'straight'? Taking the clue that light is the only standard gauge, perhaps light paths can be used. Indeed this is appropriate whenever gravity is absent, which is tantamount to our earlier observations that the 'free' test particles within the frame are removed from all extraneous matter. With gravity present the definition of an inertial frame becomes so problematic that it must be abandoned altogether, in favour of arbitrary covariant frames. In that case the concept that light defines the best possible 'straight' lines within an arena called space–time overrides our prior notions of 'straight'. Such considerations are the basis of the relativistic description of gravity, or general relativity.

5.5 Comparing Transverse Distances Between Frames

We want to compare measurements of space and time separations of events recorded in different inertial frames with each other. The starting point is to fix on an orientation between the frames as shown in Figure 5.5. The orthogonal frame O identifies the laboratory frame whilst the rocket frame, O', is similarly oriented and moves along the positive x direction. We

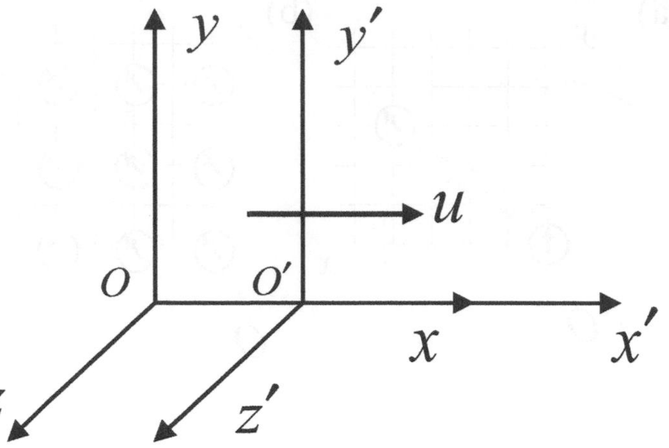

Figure 5.5 Standard orientation of frames in special relativity.

agree that the clocks at the origin of both the laboratory and rocket frames are started when the two origins coincide. This configuration of the two frames and choice of time origin is known as the **standard orientation**. Obviously we could arrange things in a more general way, but it would be much more complicated and not very interesting. We will stick with the standard orientation.

As a first step towards comparisons between frames, imagine that O and O' place metre rulers along their respective y-axes. Now O' will obviously regard his own rule as being 1 m long, but how long will he observe O's rule to be when they pass each other at $t = t' = 0$, just before the situation depicted in Figure 5.6? Although the answer may seem obvious, it is not a 'silly' question because we will see that lengths along the direction of travel are different between frames. Answer: it must be 1 m by the principle of relativity! Anything else would not only give us a mechanism for distinguishing between inertial frames, but would also lead to a contradiction. If O''s stick was shorter than O's as indicated in Figure 5.7(a) then, as they pass, O' can paint a mark on O's stick indicating how much shorter it is. However, by the principle of relativity, O's view *must* be as shown in Figure 5.7(b). He can paint a mark on O''s stick indicating how much O's stick is shorter than O''s. The two sticks are therefore *simultaneously* observed to be shorter than each other which is a contradiction.[2] *Conclusion: distances transverse to the direction of relative motion between frames are invariant*:

$$y = y', \quad z = z'. \tag{5.7}$$

These equations are actually part of the full Lorentz transformations between two frames in the standard orientation. The other two are more complicated and will be derived once we have carefully studied the behaviour of time between inertial frames.

[2]The word 'simultaneously' in this sentence is absolutely vital to the argument, and is the clue to why the same logic fails for distances along the direction of relative motion.

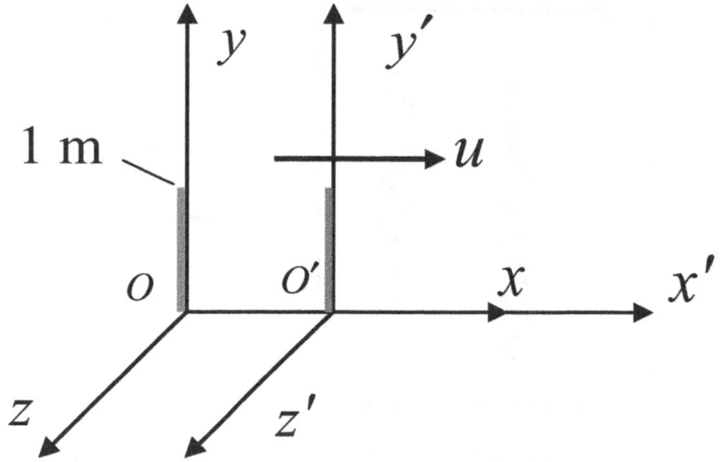

Figure 5.6 Comparison of metre sticks placed along the y-axis. Location of frames just after $t = t' = 0$.

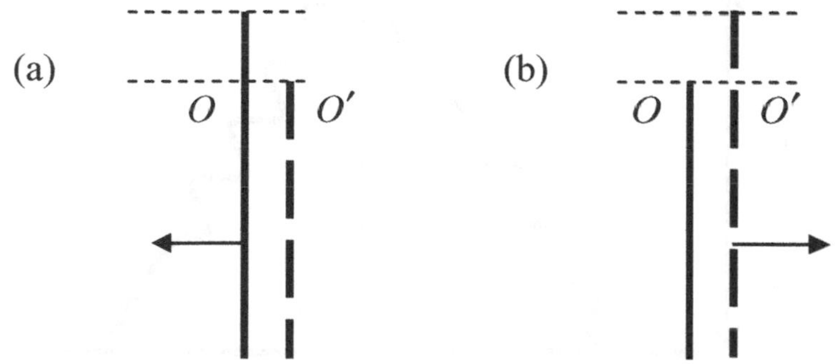

Figure 5.7 Different views of metre sticks passing each other: (a) O''s view, (b) O's view.

5.6 Lessons from a Light Clock: Time Dilation

We have seen how clocks are synchronised in a frame, but what about the construction of the clocks themselves? Since everything must be related to light, let the clock be based on the round trip of a light pulse (see Figure 5.8). The clock, at rest in O', consists of two mirrors separated along the y'-axis by a distance L'; the clock period is defined by the round trip of a light pulse leaving the bottom mirror (Event 1), being reflected at the top (Event 2 = 'tick') and finally returning to the lower mirror (Event 3 = 'tock'). If this complete cycle takes τ seconds according to O' then evidently

$$L' = c\tau/2 \quad . \tag{5.8}$$

How will this sequence of events appear from O's viewpoint (see Figure 5.9)? Since the clock

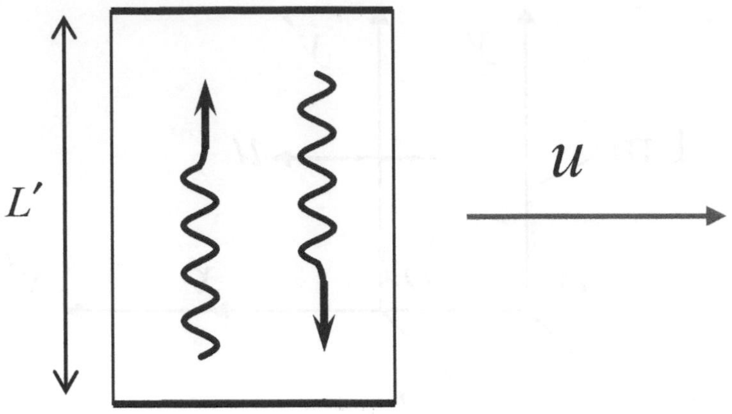

Figure 5.8 A light clock at rest in O'.

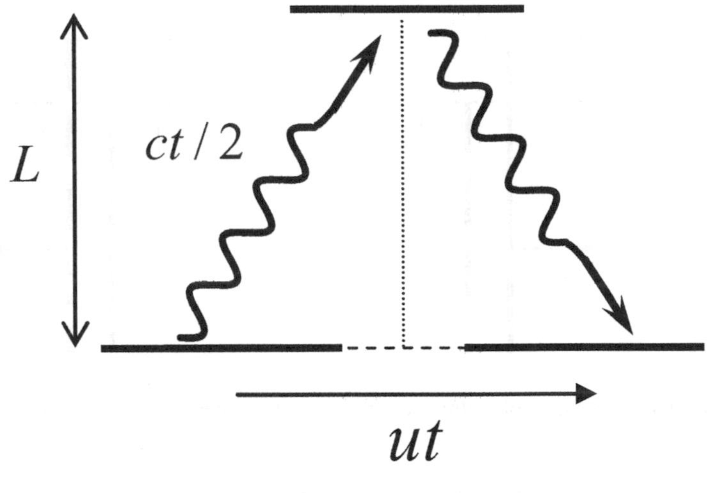

Figure 5.9 O's view of the light clock at rest in O'.

is moving along the x-axis with speed u, the light follows the 'triangular' trajectory shown. If the period of the clock measured by O is t, then the distance moved by the light pulse during the 'tick' phase is $ct/2$ and the distance moved by the clock along the x-axis during this time is $ut/2$. By applying Pythagoras' theorem to the right-angled triangle in the figure we deduce that the transverse distance, L, is given by

$$L = \sqrt{c^2 \left(\frac{t}{2}\right)^2 - u^2 \left(\frac{t}{2}\right)^2} = \frac{t}{2}\sqrt{c^2 - u^2}. \tag{5.9}$$

However, we proved in the last section that perpendicular distances are invariant so $L = L'$ and consequently

$$c\frac{\tau}{2} = \frac{t}{2}\sqrt{c^2 - u^2}. \tag{5.10}$$

Finally we deduce that the connection between t and τ is given by

$$t = \frac{\tau}{\sqrt{1 - \frac{u^2}{c^2}}} = \gamma\tau \tag{5.11}$$

where γ is the factor which was introduced in Equation (5.4). The time taken for the clock period measured by O is *longer* than the period measured by O'. This must be the case since the speed of light is the same in both frames and the light travels further in O. Indeed, *any* frame in which the light clock is not stationary will record a longer time than τ, the period recorded in a frame wherein the clock is stationary. Do not be tempted to think that this time dilation is an artefact, or simply a particular case of the light clock. *All* processes associated with bodies that are stationary in O', mechanical, biological, nuclear, etc., must be observed to occur at a slower rate when observed from a frame in which the bodies are moving. If this were not so then the light clock would become out of sync with other processes in that frame. By noting these discrepancies it would then be possible to distinguish between the frames, contrary to the principle of relativity. There is no way out, and we have deduced a fundamental property about how time transforms between frames.

The other thing to notice about Equation (5.11) is that the rocket light clock would be perceived by the lab observer to tick at an *imaginary* rate if the frame speed u were greater than c, and this already suggests that frame changes with $u > c$ are not possible.

Does all this really work in practice? Is there any simple experiment that can be performed that can confirm the concepts of time dilation? Consider the radioactive decay of moving particles. If the characteristic decay time is τ seconds in the rest frame of the nucleus then we would expect, by the above arguments, this characteristic time to be longer in a frame in which the nucleus is moving. This could be measured by counting the proportion of such particles that actually decay. A muon naturally decays into a pion and an electron such that after $t\,\mu$s the fraction of the original muon population surviving is $\exp(-t/2.2)$. Muons are created in the atmosphere at a height of about 2km as a result of the interaction of cosmic rays with nuclei, moving towards the ground with speed $u = 0.995c$. They therefore reach the ground after $6.7\,\mu$s and, without time dilation, we might expect a fraction $\exp(-6.7/2.2) \approx 0.05$ of the original number to survive. In fact, the fraction at ground level is observed to be nearer to 0.7. The explanation is that the lifetime of the muon in the Earth frame is γ times longer than in the rest frame of the muon. Here $\gamma = (1 - u^2/c^2)^{-1/2} \approx 10$, so that the surviving fraction once time dilation is taken into account is $\exp[-6.7/(10 \times 2.2)] \approx 0.7$.

5.7 Proper Time

Students are often uncomfortable with the notion that according to the lab observer the rocket clock ticks at a slower rate than a similar light clock at rest in the lab. Why, they reason, by the principle of relativity, does not the rocket observer see the lab light clock running γ times slower than the light clock at rest in his frame? The answer is that he does! However, notice that the respective observations relate to a *different set of events*. The return of the light pulse of the lab light clock is a distinct event to the pulse return of the rocket light clock pulse. These events are spatially separated along the line of relative frame motion and will not be simultaneous in *either* frame. Moreover the temporal ordering of the two events is reversed as we move from lab frame to rocket frame and vice versa. *But neither event can influence the other*. If the spaceman in the rocket sends a signal back to the laboratory saying 'My clock now reads τ', that signal will be received at the lab origin *after* his own clock has passed τ, and the latter will not think anything is untoward. A similar conclusion is reached by the person in the rocket if the person in the lab sends a signal confirming the completion of one period of the lab clock. Some details of this argument are developed in Problem 5.6. For now, we note that if a light clock ticks at a certain rate in its rest frame, then a frame in relative motion will see it tick slower. For this reason the time associated with a body at rest is the 'special' time associated with that body and it is given the name **proper time**:

- The **proper time** elapsed between two events is the time recorded by an observer for whom the two events occur at the same place.

5.8 Interval Invariance

In Section 5.6 we used the invariance of transverse distances to derive the formula for time dilation. You may have noticed then that there was a certain flexibility in the choice of speed of the rocket, for although different relative frame speeds would lead to different γ factors (since γ depends on u), for every choice of u the transverse distance is invariant. In this section we utilise this flexibility to derive an important invariant quantity. In addition to O and O' let us introduce a third frame, O'' = 'hyper-rocket frame', moving at speed $\bar{u}$ relative to O (where $\bar{u} > u$), and analyse the space–time coordinates of Events 1 (= emission of light pulse), 2 (= 'tick') and 3 (= 'tock') for all three frames. In O'' the rocket light clock is seen to proceed to the left and his view of the three events is shown in Figure 5.10. In O' the time separation between Events 1 and 3 is the clock period τ, whilst their spatial separation is zero since the light clock is at rest in the rocket frame. For the other two frames let:

Δt = time separation between Events 1 and 3 as observed by O,
Δx = space separation between Events 1 and 3 as observed by O,
$\Delta t''$ = time separation between Events 1 and 3 as observed by O'',
$\Delta x''$ = space separation between Events 1 and 3 as observed by O''.

Now the transverse distance between Events 1 and 2 is the same for all frames, or $L = L' = L''$, which yields on applying Pythagoras:

$$\left[\left(\frac{c\Delta t}{2}\right)^2 - \left(\frac{\Delta x}{2}\right)^2\right]^{1/2} = \frac{c\tau}{2} = \left[\left(\frac{c\Delta t''}{2}\right)^2 - \left(\frac{\Delta x''}{2}\right)^2\right]^{1/2}, \tag{5.12}$$

or

$$(c\Delta t)^2 - (\Delta x)^2 = (c\Delta t'')^2 - (\Delta x'')^2 = c^2\tau^2 , \tag{5.13}$$

that is the quantity $(c\Delta t)^2 - (\Delta x)^2$ is an *invariant* between frames. It is the square of a quantity called the space–time interval, which in three dimensions is defined as

$$(\text{interval})^2 \equiv (c\Delta t)^2 - (\Delta x)^2 - (\Delta y)^2 - (\Delta z)^2 = (c\Delta\tau)^2 , \tag{5.14}$$

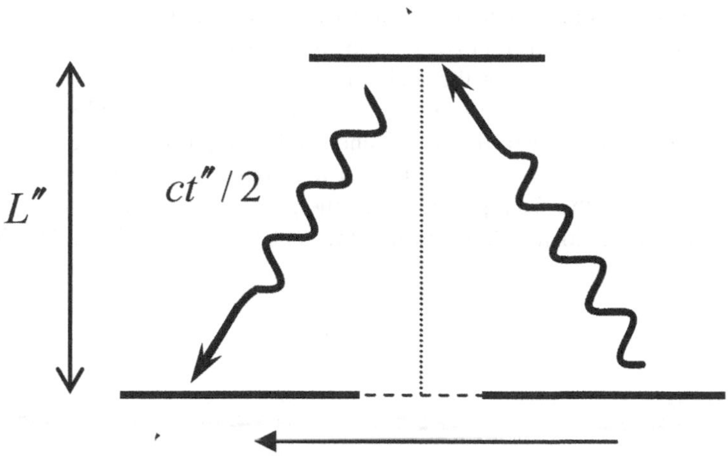

Figure 5.10 O'''s view of the light clock at rest in O'.

where $\Delta\tau$ is the proper time between the events. The result is true for *any* pair of events, not just those identified in this light clock example. Indeed, for events that are almost simultaneous ($\Delta t \approx 0$) but spatially separated ($\Delta x, \Delta y, \Delta z \neq 0$), the quantity $(c\Delta t)^2 - (\Delta x)^2 - (\Delta y)^2 - (\Delta z)^2$ is negative in which case, by analogy with Equation (5.14), it is natural to define

$$(\Delta s)^2 = (\Delta x)^2 - (\Delta y)^2 - (\Delta z)^2 - (c\Delta t)^2 \ , \tag{5.15}$$

and the resultant invariant interval is then referred to as **spacelike**. Similarly, events for which $(\Delta\tau)^2$ in Equation (5.14) is positive are said to be separated by a **timelike** interval. Events separated by a null interval are connected by a light signal travelling from the first event to the second and are therefore referred to as being separated by a **lightlike** interval.

The space–time interval plays a role very similar to distance in Euclidean geometry, and indeed Equations (5.14) and (5.15) look very similar to Pythagoras' theorem. Just as the square of the Euclidean distance between two points $(\Delta x^2 + \Delta y^2 + \Delta z^2)$ is an invariant between Cartesian frames, so the space–time interval is invariant between inertial frames in relative motion. This is very suggestive that Equation (5.14) should be regarded as defining a kind of 'distance' in space–time. However, the fact that the square of the space–time 'distance' can be positive, zero or negative distinguishes it totally from the geometry of Euclid. Nevertheless, many of the features of the new geometry in space–time are Euclidean analogues.

5.9 The Relativity of Simultaneity

We have already come a long way. The crucial features of special relativity are summed up by the light clock analysis, the fundamental features of which are all encapsulated in Figs. 5.8, 5.9 and 5.10. We have been able to derive the time dilation effect (Equation (5.11)) and interval invariance. Before moving on, there are two more effects that can also be deduced using the ideas that have been assembled so far. In this section we will discuss how events that are simultaneous but spatially separated in a particular frame are *anything but* simultaneous when viewed from a frame in relative motion to the first. Consider the situation depicted in Figure 5.11 wherein the rear of the rocket is located at the origin of O', which is in the standard orientation with respect to O. At $t = 0$ two firecrackers explode in the lab at the locations of the front and rear of the rocket and send light pulses towards each other. Since the frame origins coincide at $t = t' = 0$ the rear light pulse is emitted at $t' = 0$ in the rocket frame. Now what about the front pulse? When the pulses meet they must do so towards the rear of the rocket since during their journey the rocket moves to the right. Now since the

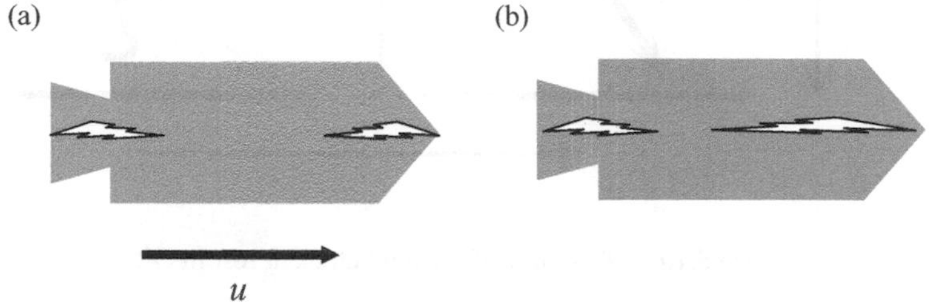

Figure 5.11 Two views of a pair of firecrackers exploding in the lab: (a) as seen from O in which the rocket is moving with velocity u, (b) as seen from O'.

front pulse has further to travel in the rocket frame, *it must have been emitted before the rear pulse*! The two spatially separated events that are simultaneous in the lab frame are *not* simultaneous in the rocket frame. Now if you have successfully got your head round this reasoning, try working out the reverse situation: two firecrackers are emitted simultaneously in the rocket frame. Can you deduce that in the lab frame the *rear* pulse is emitted first?

Now we consider the relativity of length.

5.10 The Relativity of Length: Length Contraction

When we spoke about constructing an inertial frame in Section 5.4 we assembled a latticework of rods against which the length of extended bodies at rest in the frame could be measured. We can, at our leisure, record the spatial locations of the front and rear of the body and work out its length. Thus far no problem; but when we consider the length of a moving body we have to be more careful. A consistent definition of length of a moving body is to measure the locations of its front and rear ends *at a particular time*. If the body's length is oriented perpendicular to its motion again there is no problem because transverse lengths are invariant as we showed in Section 5.5. However, what if the body's length is oriented along the line of motion? Since the simultaneous recording of the locations of the front and rear of the body will be anything but simultaneous in the body's rest frame; we anticipate that the length will be different from the body's length in its rest frame. Another method of measuring the length of a body is illustrated in Figure 5.12. A rod at rest in the lab frame is observed to pass over the rocket origin, the passage of the rear and front ends being labelled

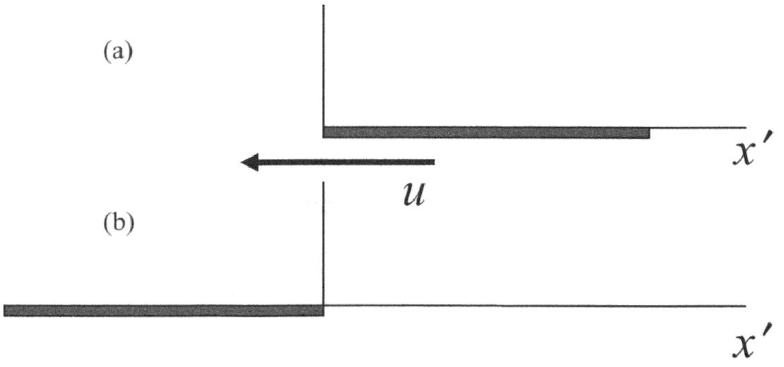

Figure 5.12 Rod at rest in the lab observed from the rocket: (a) Event 1: the rear of the rod coincides with the rocket origin, (b) Event 2: the front of the rod coincides with the rocket origin.

Events 1 and 2 respectively. By recording $\Delta t'$, the time between these events, the length of the rod in the rocket frame can be deduced from

$$l = u\Delta t'. \tag{5.16}$$

Notice that measuring the length in this way obviates the difficulties associated with making simultaneous measurements of the locations of the front and rear of the rod, since the length is deduced from two events which occur *at the same place* in the rocket frame. As far as the lab observer is concerned the two events are identified by the rocket origin passing over first the rear, and then the front of the rod as shown in Figure 5.13. The time separation between these events is recorded by the lab observer as Δt. Since the rocket origin moves with velocity u, then using the same principle, the length recorded by the lab observer is

$$l_0 = u\Delta t. \tag{5.17}$$

The subscript on l_0 reminds us that this is the rest length (or **proper length**) of the rod; but we know from the light clock analysis that since the two events occur at the same place in the rocket frame $\Delta t = \gamma\Delta t'$. Combining this with Equations (5.16) and (5.17) we obtain

$$l = (1 - u^2/c^2)^{1/2} l_0 = \frac{l_0}{\gamma}\,. \tag{5.18}$$

In the rocket frame the rod appears shorter than it appears in its rest frame! This feature is referred to as **length contraction**. Notice that although we obviated the difficulty of measuring spatially separated events in the rocket frame by locating the events associated with the measuring process at the rocket origin, the two events used *are* spatially separated in the lab frame.

5.11 The Lorentz Transformations

We have examined a variety of special cases in which the space and time separation of events have been related to each other in different frames. We must now devise a general

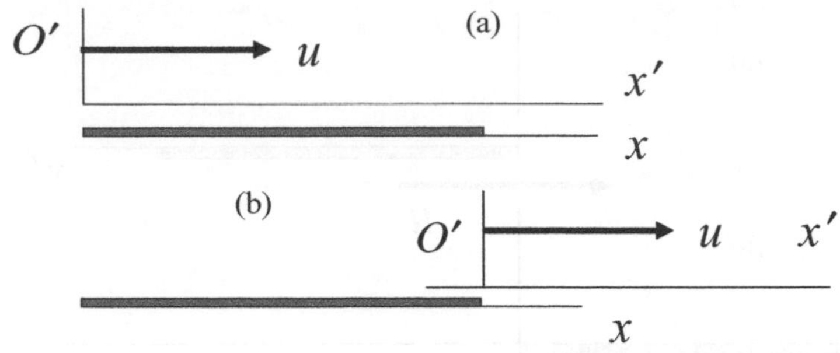

Figure 5.13 Passage of the rocket origin over the rod at rest in the lab: (a) Event 1: origin passes over the rear of the rod, (b) Event 2: origin passes over the front of the rod.

prescription, or in other words, if two events are separated in space and time by $(\Delta x, \Delta t)$ in the lab frame, what are the corresponding quantities $(\Delta x', \Delta t')$ in the rocket frame? Notice how geometrical this problem sounds. In Euclidean geometry we typically want to transform between coordinate systems that are rotated with respect to each other. Here we wish to transform space–time coordinates, the resultant prescription being known as the **Lorentz transformation**. The method we will follow is exactly the one Einstein used.

Starting again with the standard orientation of Figure 5.5 (reproduced here as Figure 5.14) let us choose one of the events to be coincident with the origins of O and O' at $t = t' = 0$. This is purely a matter of where we choose to locate coordinate origins, and when we choose to start the clocks. The other event has coordinates (x, y, z, t) in O and (x', y', z', t') in O'. Setting it up in this way means that we can now drop the Δ from the notation. Transverse

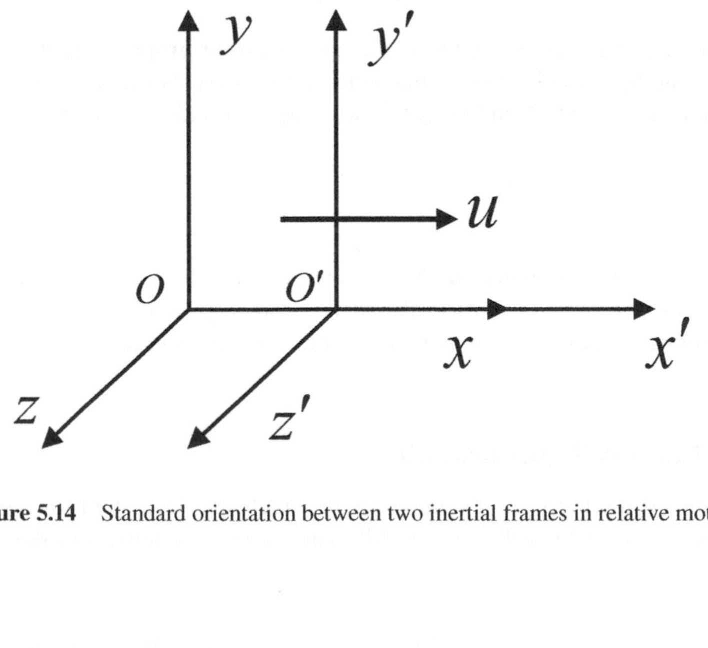

Figure 5.14 Standard orientation between two inertial frames in relative motion.

distances are invariant so immediately we have

$$y = y' , \tag{5.19}$$

$$z = z' . \tag{5.20}$$

Suppose that the remaining quantities are related linearly through

$$x' = Ax + B\,(ct) \ , \tag{5.21}$$

$$t' = D\left(\frac{x}{c}\right) + Et, \tag{5.22}$$

where the coefficients (A, B, D, E) are to be determined, and the factors of c are introduced for convenience to make the these coefficients dimensionless. What right have we to presume a linear relationship, and not one that includes terms like x^2 or $\sin(t)$ on the right-hand sides? One objection is that a quadratic relationship might yield multiple solutions when we try to invert it, and certainly to have a single event with two time coordinates clearly will not do![3] Another objection might be that when we come to invert the transformation (that is expressing (x, t) in terms of (x', t')) the inverse transformation looks different from the original (e.g. $x' = \sin^{-1}(x)$ is distinct from $x = \sin(x)$), and this would be forbidden by the principle of relativity which says that physics must look the same in all inertial frames. (The $x' = \sin(x)$ example is also forbidden on dimensional grounds.) However, we cannot totally exclude the possibility of a more general transformation for which these objections are not valid, although the restrictions they impose are severe. Ultimately there is no 'proof' that the relationship must be linear, but this is certainly the simplest presumption to make. What we can assert is that *if* the relationship is linear then Equations (5.21) and (5.22) represent the most general form possible.[4]

Dividing Equation (5.21) by (5.22) yields

$$\frac{x'}{t'} = \frac{A\frac{x}{t} + Bc}{D\frac{x}{ct} + E} . \tag{5.23}$$

Now a light pulse (as always!) which is travelling in the positive x direction will move with velocity $+c$ in both frames so $x/t = x'/t' = c$. Equation (5.23) becomes, after cancelling off the factors of c:

$$1 = \frac{A + B}{D + E}. \tag{5.24}$$

Similarly, a pulse travelling in the negative x direction yields:

$$-1 = \frac{-A + B}{-D + E} . \tag{5.25}$$

Finally we know that the origin of O moves at velocity $-u$ with respect to the origin of O', so that $x'/t' = -u$ when $x/t = 0$, so

$$-\frac{u}{c} = \frac{B}{E} . \tag{5.26}$$

[3]Technically, the name for transformations which yield single values both for the direct transformation and for the inverse is a *bijection*.

[4]Since the relationship must be invertible it is possible to show that this places an additional constraint on the undetermined coefficients, namely $AE - BD \neq 0$, though we will not be using this.

Now as an exercise you can show that Equations (5.24) to (5.26) can be combined to give

$$E = A; \quad B = D = -\frac{u}{c}A. \tag{5.27}$$

We are nearly done. The coefficients B, D and E have been determined in terms of A, so if we can just figure out A then we would have everything. The final piece of information is to use interval invariance:

$$(ct')^2 - x'^2 \equiv (ct)^2 - x^2 . \tag{5.28}$$

Once Equations (5.21) and (5.22) are substituted into (5.28) we have an identity, true for all x and t, which helps us fix A. Setting $t = 0$, for example, gives

$$D^2 - A^2 = -1 . \tag{5.29}$$

(Just in case you are wondering, setting $x = 0$ does not yield any new information.) Combining this with Equation (5.27) gives

$$A = E = \frac{1}{\sqrt{1 - (u/c)^2}} \equiv \gamma , \quad B = D = -\frac{u}{c}\frac{1}{\sqrt{1 - (u/c)^2}} = -\gamma\frac{u}{c} , \tag{5.30}$$

and the derivation is complete. For reference we reproduce the **Lorentz transformation** (LT) equations here:

$$\begin{aligned} t' &= \gamma\left(t - \tfrac{u}{c^2}x\right) , \quad & x' &= \gamma\left(x - ut\right) , \\ y' &= y , & z' &= z . \end{aligned} \tag{5.31}$$

These equations enable us to calculate the space and time coordinates of an event (with respect to a reference event at the mutual origin of space and time in the standard orientation) in the rocket frame, given the coordinates in the lab frame. Of course if we are given the coordinates in the lab frame then we need to calculate everything the other way round. This can be done either by algebraically inverting Equations (5.31), or by simply interchanging the primed for unprimed quantities and setting $u \to -u$. Either way the result is

$$\begin{aligned} t &= \gamma\left(t' + \tfrac{u}{c^2}x'\right) , \quad & x &= \gamma\left(x' + ut'\right) , \\ y &= y' , & z &= z' , \end{aligned} \tag{5.32}$$

which are collectively known as the **inverse Lorentz transformation** (ILT) equations. Which transformation you use in a given problem depends on the data given.

When the frame velocity is small $\gamma \approx 1$ and the first two equations of the LT take the approximate form

$$t' = t - \frac{u}{c^2}x , x' = x - ut . \tag{5.33}$$

Newton's absolute time, in which $t' = t$ for any pair of inertial frames, is only recovered in this limit if in addition $|ux/c^2| \ll t$. It is therefore possible for relativistic effects to be significant even at low velocities, provided the spatial separation of the events under consideration is sufficiently large. This issue is discussed in Problem 5.10.

5.12 Velocity Addition

Now we can derive Equation (5.1). The problem, illustrated in Figure 5.15, is to calculate the velocity, $v = dx/dt$, of a body which is moving along the x'-axis with velocity $v' = dx'/dt'$. In other words, how do velocities transform between frames? We will only consider velocities that are parallel to the relative frame velocity. Velocity is a relationship between incremental changes in space and time, so we had better take the differential of the ILT, Equation (5.32) (inverse because we are given rocket data):

$$\begin{aligned} dx &= \gamma\left(dx' + u dt'\right) , \\ dt &= \gamma\left(dt' + \tfrac{u}{c^2} dx'\right) . \end{aligned} \tag{5.34}$$

Dividing these two equations and noting that $v = dx/dt$ and $v' = dx'/dt'$ we obtain

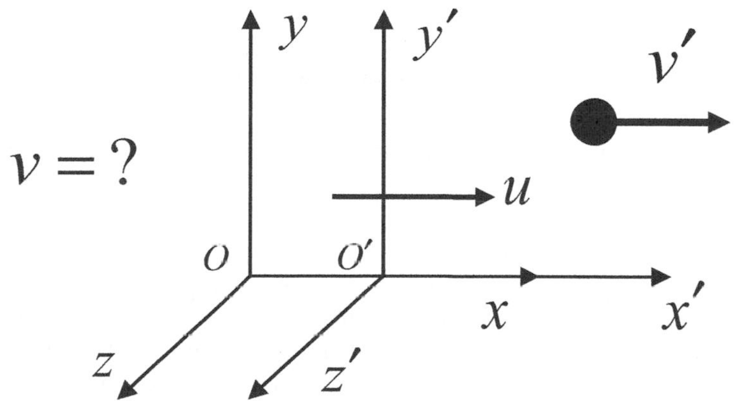

Figure 5.15 Velocity addition. What is v, given v' and u?

$$v = \frac{u + v'}{1 + {uv'}/{c^2}} . \tag{5.35}$$

It is as easy as that. Notice that this produces the familiar $v = u + v'$ when $uv'/c^2 \ll 1$ or when ***either*** u ***or*** v is much less than the speed of light. Also if u ***and/or*** v' equals c then $v = c$ also. Adding an increment to either the frame speed or the rocket particle speed takes the lab particle speed closer to, but never in excess of, the speed of light. A simple example illustrates

Example: An object moves with velocity $0.8c$ in O' which is moving with velocity $0.5c$ relative to O. What is the velocity of the object in O?

Answer: $v' = 0.8c$ and $u = 0.5c$, so using Equation (5.35) $v = 0.93c$.

Velocity transformation is illustrated in Figure 5.16 which plots v against v' for various frame velocities u. $u = 0$ is a straight line at 45° since $v = v'$, and $u = c$ is the straight line $v = c$.

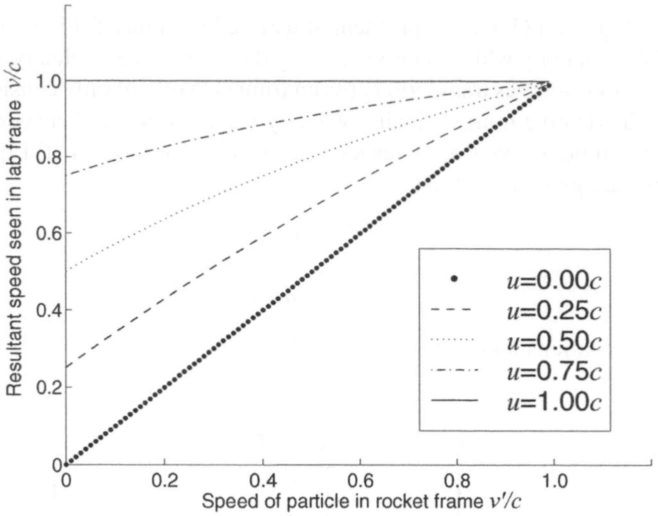

Figure 5.16 Velocity addition. Plots of v against v' for various frame velocities u.

All the other lines are not straight. One important feature concerning velocity addition is worth emphasising. In Figure 5.15 we are dealing with a single body being observed from two different inertial frames. A completely different situation is illustrated in Figure 5.17, which shows two bodies observed from a single reference frame. If the initial separation of the bodies is L and $v_1 > v_2$, how long is it before the faster body catches up the slower one? Remember to think of the question in terms of the data recorded by the clocks located in the latticework of the lab frame. From this data the relative rate at which the gap closes is $(v_1 - v_2)$ so that the convergence time is $L/(v_1 - v_2)$. In this case the velocity combination appears to revert to its 'normal' value. However, this relative velocity is 'geometrical' and does not refer to the actual physical velocity of a single particle within a frame. No particle actually moves at this relative speed. Indeed, such a relative velocity can exceed the speed of light. When two photons fly apart in the lab for example, their rate of separation according to the lab observer is $2c$. The attempt to construct a 'genuine' velocity which is faster than light is considered next.

5.13 Particles Moving Faster than Light: Tachyons

The principle of relativity has led us to the correct way of relating the space–time coordinates of events in one frame, to those of another frame, that is the Lorentz transformations. Regarding the relative frame velocity u as a parameter, it is clear that the LTs are well defined as u increases from zero and approaches c. At $u = c$ the LT is singular (since γ is then infinite) and after that we have the equally incomprehensible result that the space and time coordinates of events in the rocket frame become imaginary with respect to their coordinates in the lab frame (since γ is then purely imaginary). So it appears that once space–time coordinates of

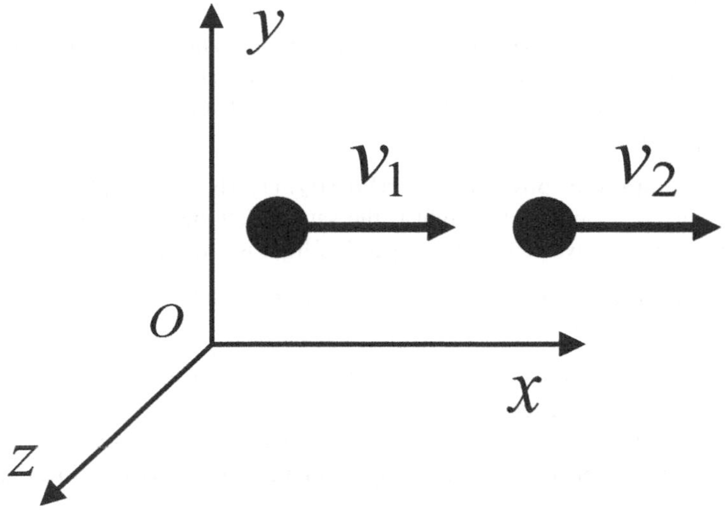

Figure 5.17 Two bodies moving in the lab frame.

an event are given with respect to one inertial frame, that event cannot be viewed from a frame moving with relative velocity equal to or greater than the speed of light. Interestingly, the velocity transformation Equation (5.35) is still well defined for $u > c$ as the γ factors cancelled out in its derivation. A particle at rest (that is $v' = 0$) in a rocket frame moving with speed $u > c$ will, according to Equation (5.35), appear to the lab observer to be moving with speed $v = u > c$. You might object that in this case a faster-than-light particle was produced via a frame change for which the γ factor is imaginary. So a more pointed question is 'Can a material particle *ab initio* travel faster than light in a given frame?' Suppose that such a putative particle, called a **tachyon**, is moving in the lab frame with velocity $w > c$. It is emitted from the lab origin and after travelling for time t it reaches position $x = wt$. Here it causes an explosion which kills an unfortunate bystander. Now according to a rocket observer moving with velocity $u(< c)$ with respect to the lab, this event will occur at time

$$t' = \left[\frac{1 - uw/c^2}{\sqrt{1 - u^2/c^2}}\right] t\,, \tag{5.36}$$

which is just the result of inserting the event with space–time coordinates $(x = wt, t)$ into the first of the LT equations (Equation (5.31)). Now it is possible for u to be greater than c^2/w for this is still $< c$. For rocket frames travelling faster than c^2/w, $t' < 0$ and in these frames the death of the bystander occurs *before* the tachyon was fired! Most physicists regard such a breakdown of causality as unacceptable on physical grounds, and the result is interpreted as indicating that no material particle can travel faster than light.

5.14 Summary

Section 5.2. Galilean relativity: Mechanics experiments cannot distinguish between inertial frames. **Invariant** quantities take the same value in different inertial frames. **Covariant** quantities take distinct values in different inertial frames, but in such a way as to maintain the same form of physical laws.

Section 5.3. Fundamental postulates of relativity: (1) The laws of physics are the same in all inertial frames; (2) the speed of light is the same in all inertial frames. Equivalently: (1) the laws of physics are covariant; (2) the speed of light is invariant.

Section 5.4. 'Observer' = orthogonal latticework of rods and synchronised clocks recording events in an inertial frame.

Section 5.5. Compare times and distances between events using the **standard orientation** (Figure 5.5) in which the rocket frame O' moves with velocity u with respect to the lab frame O with their respective x-axes aligned. Take the coincidence of the origins as the time origin for both frames. By the principle of relativity, **transverse distances** are invariant:

$$y = y', \quad z = z' \quad . \tag{5.7}$$

Section 5.6. A **light clock** consists of a light pulse travelling between two mirrors (Figure 5.8). On account of (i) the light having further to travel in a frame where the clock is moving (Figure 5.9) and (ii) the invariance of transverse distances, the time taken for the light round trip is γ times longer than the time taken in the clock's rest frame:

$$t = \frac{\tau}{\sqrt{1 - \frac{u^2}{c^2}}} = \gamma\tau \ . \tag{5.11}$$

This dilation is true for all processes, not just those associated with light.

Section 5.7. τ is the **proper time** associated with the light clock.

Section 5.8. For any pair of events separated in space and time with coordinates $\Delta t, \Delta x, \Delta y, \Delta z$, the quantity

$$(c\Delta t)^2 - (\Delta x)^2 - (\Delta y)^2 - (\Delta z)^2 \ , \tag{5.14}$$

is invariant and is called the square of the space–time **interval**. If this quantity is positive/negative/zero, the events are respectively referred to as having **timelike/spacelike/lightlike** space–time separation.

Section 5.11. The general prescription for relating the space–time separation between a pair of events in one inertial frame to those of another in relative motion (in the standard orientation, with a reference event located at the coincidence of the origins) is given by the **Lorentz transformation** (LT) equations:

$$\begin{aligned} t' &= \gamma\left(t - \tfrac{u}{c^2}x\right) , \quad & x' &= \gamma\left(x - ut\right) , \\ y' &= y , & z' &= z , \end{aligned} \tag{5.31}$$

or by the **inverse Lorentz transformation** (ILT):

$$t = \gamma\left(t' + \tfrac{u}{c^2}x'\right)\,, \qquad x = \gamma\left(x' + ut'\right)\,,$$
$$y = y'\,, \qquad z = z'\,. \tag{5.32}$$

Consequences:

- Events that are simultaneous in one frame (e.g. $t = 0$), are not simultaneous ($t' \neq 0$) in another frame if they occur at different positions along the axis defined by the relative frame motion; the **relativity of simultaneity, Section 5.9**.
- A ruler at rest in the lab is measured in the rocket frame to have length l_0/γ where l_0 is the ruler's rest length; **length contraction, Section 5.10**.

Section 5.12. Differentiating the ILT leads to the velocity addition formula

$$v = \frac{u + v'}{1 + {uv'}/{c^2}}\,, \tag{5.35}$$

where v' is the given velocity of a body along the x'-axis of the rocket frame.

5.15 Problems

5.1 At what speed is a metre rule moving relative to an observer who finds its length only 750 mm?

5.2 According to Mr Spock's clock, Captain Kirk's clock is running slow by 5%. What is their relative speed?

5.3 Plot a graph of the Lorentz factor, $\gamma = (1 - u^2/c^2)^{-1/2}$, as a function of u/c from 0 to 1 in steps of 0.1. Above what speed are relativistic corrections to Galilean kinematics more than 10%?

5.4 Use the binomial expansion to find an approximation for γ when $u \ll c$, up to terms of order $(u/c)^2$.

5.5 Alpha Centauri is 4.4 light years away (as measured in the Earth's frame of reference). If an astronaut wants to reach it in 10 years as reckoned by his own clock, at what (constant) speed would he need to travel? How long would the trip take according to his twin brother left behind on the Earth?

5.6 A student complained after one of my lectures that the theory of relativity must be wrong because using the theory one can argue as follows: 'a light clock at rest in the rocket frame O' is seen to go slow by an observer at rest in the laboratory frame O. Similarly a light clock at rest in O is observed by O' to go slow. Both clocks cannot both be "going slow", hence contradiction.' By considering the space–time coordinates of the four events: (1) light pulses emitted by clock at rest in O and O' (by choice of space–time origin in the standard orientation these two events have the same coordinates, and may therefore be regarded as the same event), (2) light clock at rest in O completes one period, (3) light clock at rest in O'completes one period, and (4) the event simultaneous with (3) in the lab frame of reference, but located at the lab origin, show how the paradox is resolved.

5.7 *Hit or miss?* Captain Kirk (K) and Mr Spock (S) pilot identical spaceships of length L (in their own respective rest frames):

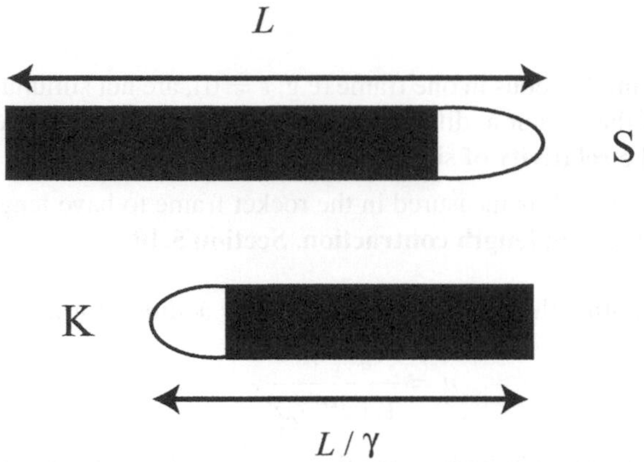

Figure 5.18 Figure for Problem 5.7.

A prearranged practice session involves K and S approaching each other almost head-on, with high relative speed, u. At the instant when the *nose* of S's ship is level with the *tail* of K's ship then S will fire a warning shot from a gun in his tail in front of K (see Figure 5.18). S has reassured K that this is harmless, due to the Lorentz contraction of K's ship to a length L/γ as illustrated. Imagine K's worried expression when approaching head-on and observing S's ship contracted! Ignore the transverse separation of the ships and bullet travel times. Does the bullet hit or miss? Analyse using Lorentz transformations the problem from both Kirk's and Spock's viewpoint.

5.8 In Problem 5.5, you calculated the speed necessary to reach Alpha Centauri (distance 4.4 light years from Earth) in 10 years as measured by the rocket clocks. Presume that on reaching Alpha Centauri the travelling twin immediately turns around and returns home at the same speed. How much younger will the travelling twin be on his return to Earth with respect to his brother who remained at home? Why is it incorrect for the travelling twin to conclude that from his perspective the Earth has been receding, and then approaching, so that it is he, not his brother, who ages faster? (*Hint: Think carefully about what each brother feels between departure and meeting up again; does each brother have an identical history?*)

5.9 An observer in the lab frame sees two particles travelling respectively at $V = +0.8c$ and $V = -0.8c$ along the x-axis.

(a) With what velocity does an observer stationary with respect to the first particle see the second particle travel?

(b) Similarly, with what velocity does the first particle move with respect to the second?

(c) What is the relative velocity between the two particles as observed from the laboratory? (*Care*!) Comment briefly on your answer to part (c).

5.10 Two individuals S' and S'' are walking towards each other along a road each at a speed of $3\,\mathrm{m\,s^{-1}}$ relative to the road. They cross at a site occupied by a stationary third observer, S. All agree to set their time origin at the crossover point, that is $t = t' = t'' = 0$. Near a star that lies on the line of the road, 4 light years away, a spaceship at rest in the frame of S launches a missile at $t = 0$ destined to destroy the Earth some time in the future. Calculate the time when the missile launch occurs in the frames S' and S'', stating carefully in each case whether it is earlier or later than in S. Ignore the effects of gravity and ignore the rotation of the Earth. Comment on which, if any, of the Earthbound observers can actually discuss the Earth's fate when they meet.

5.11 A particle is fired from the rear end of a rocket with velocity $+v'$ as observed by someone on the rocket. The rocket travels with velocity u relative to a lab observer, who in turn sees the particle travelling with velocity $+v$. The particle then reaches the front end of the rocket where it rebounds elastically and returns to the rocket rear.

(a) If L_0 is the rest length of the rocket, show that the time recorded by the rocket observer for the particle to complete the roundtrip is given by $2L_0/v'$.

(b) If L is the length of the rocket according to the lab observer, and v_+ and t_+ are respectively the velocity and the time taken for the first part of the roundtrip as measured by the lab observer, show that $v_+t_+ + L = ut_+$.

(c) Hence show that the times recorded by the lab observer for the particle to reach the end of the rocket, (t_+), and then to return to the starting point, (t_-), are given by $t_\pm = \gamma^2 L\left(1 \pm uv'/c^2\right)/v'$.

(d) Hence by evaluating the total round trip time taken in the lab frame, prove the length contraction formula.

5.12 *This is the famous 'pole and barn' paradox.* Paul Volta has a pole 20 m long. His run up is so fast that his pole appears contracted to a length of only $L = 10\,\mathrm{m}$. To achieve such a speed he needs a long run up and he has to pass through the changing rooms, which are just over $L = 10\,\mathrm{m}$ long. In an attempt to remove him from the competition his competitors install a photocell system, designed to close both entrance and exit doors at the instant when the front end of the pole reaches the exit door, with obvious disastrous consequences. Fortunately Paul hears of the dastardly plot. Though he cannot disconnect the device or prevent the doors closing, he devises another system that rapidly reopens the doors allowing him a free run. Paul therefore feels relaxed, until he begins his run up towards the changing room, when he is dismayed to observe it has contracted. Too busy training to have attended relativity lectures he worries about losing his title, his reputation and 15 m of pole. Take as origin of coordinates and time the event R, when the front of the pole reaches the entrance door of the changing room (see Figure 5.19).

(a) From the fact that the pole appears contracted to 10 m calculate the value of γ. Hence calculate Paul's speed u.

(b) Calculate the length of the changing room according to Paul.

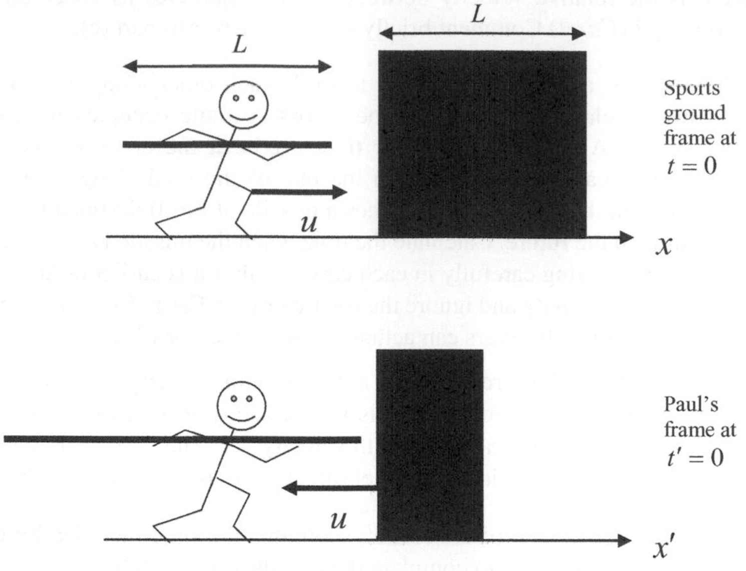

Figure 5.19 Figure for Problem 5.12.

(c) Write down expressions (in terms of L, the length of the changing room in the sports ground frame, and u, the relative speed between the frames) for the space–time coordinates in the sports ground frame of the following events: (i) the front end of Paul's pole reaches the exit door (event F), (ii) the back end of Paul's pole reaches the entrance door (event B).

(d) Use the Lorentz transformation equations to obtain expressions for space–time coordinates (in terms of L and u) of these same events in Paul's frame.

(e) Calculate the times t_F, t_B, t'_F and t'_B in nanoseconds (t' refers to time as measured in Paul's frame).

(f) Compare the values of t'_F and t'_B obtained from part (e). What can you say about Paul's chances?

(g) Draw a graph of t vs. x (that is a space–time diagram), one for each frame, of the trajectories of the entrance and exit doors, the front and back of the pole and the coordinates of the events R, F and B.

(h) For a real mind-bender, try working out the countdown to disaster if Paul's countermeasures fail and the changing room doors remain closed. (*This is very tough; the key is to recognise that relativity necessitates a revision of our everyday notions of rigidity.*)

6

Relativity 2: Energy and Momentum

Actually, we cannot measure the length of real macroscopic objects since they are hard to accelerate up to relativistic speeds, and the 'length' of small particles is ill defined due to quantum effects. Time dilation in particle decays is significant as we have seen, but again, the twin paradox of Problem 5.8, though interesting to discuss, is not a technological possibility at present. So despite its aesthetic attraction, the role of space–time dilation is rather limited in terms of how relativity affects the 'real' world of experimental physics. The real experimental arena at present is in the high-energy particle physics laboratory and there, through recording particle tracks and detecting scintillations, it is the **energy** and **momentum** of particles that are measured. It is to these quantities that we must now turn.

6.1 Energy and Momentum

If nothing can travel faster than light then, according to Newtonian theory, the momentum of a body is limited to $p \leq mc$. In itself there is nothing seriously wrong with the notion that momentum is limited, although we might be uncomfortable with the idea that as we crank up the force on a body its momentum does not increase beyond a certain value. What is more serious is that experimentally, specifically in high-energy collisions, the total system momentum

$$p_{\text{total}} = \sum_{\text{particles}} mv \,, \tag{6.1}$$

is *not* conserved for an isolated system. Nor, in fact, is the total system kinetic energy

$$K_{\text{total}} = \sum_{\text{particles}} \frac{1}{2}mv^2. \tag{6.2}$$

Recall, however, from Chapter 4, that the kinetic energy is only conserved if the interactions between bodies are 'hard', and do not dissipate energy into other forms (heat, electromagnetic, elastic, etc.). Although at the level of particle physics experiments it is harder to identify these 'alternative' sinks for energy in inelastic collisions, it may seem that violation of energy conservation is less of a problem than momentum conservation. So we have a dichotomy: either we must abandon the Newtonian expression for momentum, or we

Classical Mechanics – Second Edition From Newton to Einstein: A Modern Introduction Martin W. McCall

must redefine momentum in such a way that it *is* conserved. This cuts right through the middle of Newton's laws, for we saw in Chapter 1 that the conservation of momentum for isolated systems was an emergent property of the second law. Newtonian physics is in serious trouble if we are going to attempt to maintain the conservation of momentum for isolated systems. Now that we are faced with the possibility of abandoning Newtonian momentum, once again the issues over the role of definition and physical law, discussed in Section 1.5, rear their head again. If momentum is going to be *defined* to be a quantity that is conserved in high-speed collisions (with the additional requirement that it reduces to Newtonian momentum for low speeds) then is not our reasoning circular? No! The point is that one experiment is not enough. The first experiment can be used to define, or propose, an expression for momentum that satisfies the Newtonian limit condition and is conserved in *that* experiment. Now we can perform additional experiments to see if momentum so defined is conserved *in general*. It is in the arena of these additional experiments that we are probing nature to find out if our definition is actually something of fundamental significance.

So how to proceed? First, let us consider the **world line** of a particle on a **space–time diagram** (see Figure 6.1). This is simply a plot of the particle's trajectory, or its position as a function of time. The convention is to plot time on the vertical axis and space on the horizontal axis. We know that in the standard orientation the transverse spatial dimensions transform trivially between frames, so the x–t plane is where our interest lies. It is indeed fortunate that only two dimensions are of interest, as this is the limit of dimensionality of a piece of paper! Now the world line of the particle is just the locus of all the events it experiences joined together to make a continuous path in space–time. If we choose two nearby events on the world line characterised by a time separation dt, and a spatial separation

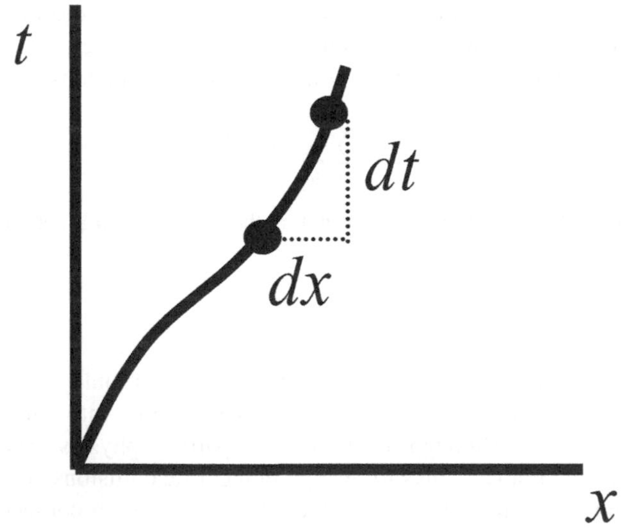

Figure 6.1 Space–time diagram of a particle trajectory.

dx, then we know from interval invariance that

$$cd\tau = \left[(cdt)^2 - (dx)^2\right]^{1/2} \tag{6.3}$$

is an invariant quantity. In this expression, $d\tau$ is just the infinitesimal proper time between the events, or the time between them observed from a frame in which the particle is at rest. (For events that are sufficiently close, the particle's velocity is approximately constant, so the rest frame is well defined, even if the particle is accelerating as in Figure 6.1.) Now multiplying Equation (6.3) by other invariant quantities produces new invariants. We will multiply by $m/d\tau$, where m is the mass of the particle measured in the rest frame, or simply the **rest mass**. Whether this is the mass measured from a frame in which the particle is moving is not an issue. The quantity m is *defined* with respect to the frame in which the particle is as rest and that number can be communicated to any other observer. The rest mass is therefore invariant by definition. Multiplying through gives

$$mc = \left[\left(mc\frac{dt}{d\tau}\right)^2 - \left(m\frac{dx}{d\tau}\right)^2\right]^{1/2} = \left[(F)^2 - (Q)^2\right]^{1/2}, \tag{6.4}$$

where just for a moment we have defined $F \equiv mcdt/d\tau$ and $Q \equiv mdx/d\tau$. Now from Equation (6.3) it is readily seen that

$$d\tau = dt\left[1 - \frac{1}{c^2}\left(\frac{dx}{dt}\right)^2\right]^{1/2} = dt\left(1 - \frac{v^2}{c^2}\right)^{1/2} = \frac{dt}{\gamma}, \tag{6.5}$$

where $v = dx/dt$ is the particle velocity, and $\gamma = (1 - v^2/c^2)^{-1/2}$ is the γ factor we had previously. Now what about F and Q? It is clear from their definitions and from Equation (6.5) that they may be written as $F \equiv \gamma mc$ and $Q = \gamma mv$, but do they carry any physical significance? We first examine the low velocity limit of $Q = \gamma mv$:

$$Q = mv\left(1 - \frac{v^2}{c^2}\right)^{-1/2} = mv\left(1 + \frac{v^2}{2c^2} + \cdots\right). \tag{6.6}$$

When $v \ll c$, we have, to lowest order in (v/c), $Q = mv$. Hence for low speeds Q reduces to the Newtonian expression for momentum and tentatively we identify $Q = \gamma mv = p$ as the *relativistic* expression for momentum. Moreover, and decisively, experiments show that the value of Q, summed over all the particles in an isolated system, is a conserved quantity. We have all the justification we need for identifying Q with the relativistic expression for momentum, p:

- The momentum, $\mathbf{p}$, of a particle of mass m, moving with velocity $\mathbf{v}$ is given by

$$\mathbf{p} = \gamma m\mathbf{v}, \tag{6.7}$$

where $\gamma \equiv (1 - v^2/c^2)^{-1/2}$ and $v \equiv |\mathbf{v}|$.

We have given the definition using vector notation, though in the case of a particle moving along the x-axis $|\mathbf{p}| = p_x \equiv p$. The comparison between the Newtonian $p = mv$ and the relativistic $p = \gamma mv$ shown in Figure 6.2 indicates their significant divergence for $v \geq 0.6c$.

Be careful to understand precisely what is meant by 'conserved'. We mean that whereas the momenta of individual bodies may change as they collide with one another, the total system momentum stays the same, provided of course the system is isolated, meaning that no external forces act on it. That is precisely what momentum conservation in Newtonian physics means.

Now Equation (6.4) becomes

$$mc = \left(F^2 - p^2\right)^{1/2}, \tag{6.8}$$

and we are in a position to discuss the meaning of F. If there are several particles present we can sum over them to form the total momentum $p_{\text{total}} = \sum_{\text{particles}} \gamma m v$. Then, by analogy with Equation (6.8), we could form

$$m_{\text{sys}} = c^{-1}\left(F_{\text{total}}^2 - p_{\text{total}}^2\right)^{1/2}, \tag{6.9}$$

where

$$F_{\text{total}} = \sum_{\text{particles}} \gamma m c\,. \tag{6.10}$$

The left-hand side of Equation (6.9) might then be considered the **total system rest mass**, m_{sys}. Whatever m_{sys} represents, it is evidently *not* the sum of the individual rest masses.

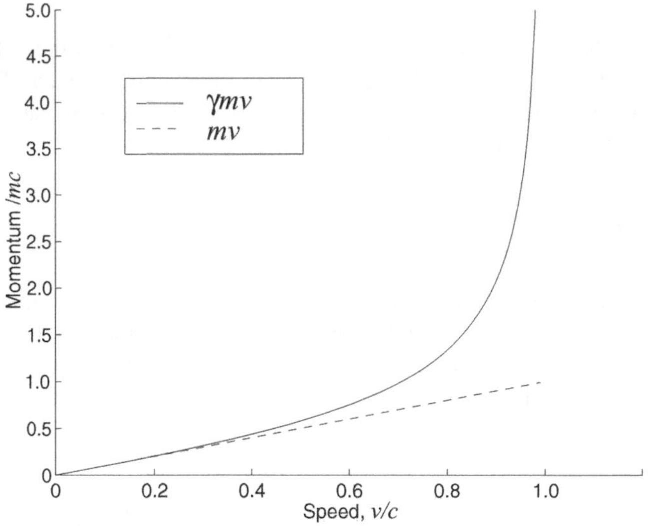

Figure 6.2 Comparison of relativistic and Newtonian expressions for momentum (cf. Equation (6.6)).

This is only the case when all the particles comprising the system are at rest. That m_{sys} does measure the total inertial mass of a system of particles is established if, in the system's rest frame, we weigh the box of particles, since, by the proportionality between inertial and gravitational mass, the weight of the box is proportional to its inertial mass. Then, by the same comments as were applied to an individual body, m_{sys} is an *invariant* between frames. A good example is a box containing a gas in the box's rest frame. The total momentum is zero, so according to Equation (6.9) the system rest mass is in this case

$$m_{\text{sys}} = \left[\sum_{\text{particles}} (\gamma m)^2\right]^{1/2} . \tag{6.11}$$

Since $\gamma \geq 1$ for all gas molecules, the system rest mass of the gas is *greater* when it is hot than when it is cold. Only in the zero temperature limit, when all gas particles are stationary ($\Rightarrow \gamma = 1$ for all particles), does the system mass equal the sum of the particle masses. We will amplify this comment shortly when interpreting F, but for now let us note the importance of the link just established between the energy of a system (any definition of energy must conform to the idea that heating a gas increases its energy content) and m_{sys}. Now experiments indicate that m_{sys} is conserved for an isolated system which is hardly surprising if we take again the box of particles paradigm – we do not expect collisions of gas molecules within the box to change its overall mass.

Now what about F? If experiments establish that m_{sys} and p_{total} are separately conserved, then Equation (6.9) enforces the conclusion that F_{total} *is also a conserved quantity of the system*. It therefore *must* be a physically important quantity, so maybe F reduces to something recognisable in the low velocity limit, like Q. Let's see:

$$F = \frac{mc}{\sqrt{1 - v^2/c^2}} = mc\left(1 - v^2/c^2\right)^{-1/2} = mc\left(1 + \frac{v^2}{2c^2} + \cdots\right) . \tag{6.12}$$

The second term ($= mv^2/2c$) is, except for a factor of c, the Newtonian expression for kinetic energy which suggests that, dimensionally at least, we identify

$$E = cF = \gamma mc^2 , \tag{6.13}$$

as the relativistic **energy** of a particle moving with velocity v:

- The energy, E, of a particle of mass m, moving with velocity $\mathbf{v}$ is given by $E = \gamma mc^2$, where $\gamma \equiv (1 - v^2/c^2)^{-1/2}$ and $v \equiv |\mathbf{v}|$.

E thus defined reduces in the low velocity limit to something *related* to the Newtonian kinetic energy

$$E \approx mc^2 + \frac{1}{2}mv^2 = mc^2 + K_{\text{Newtonian}} . \tag{6.14}$$

We already know that E_{total}, the sum of γmc^2 over all the particles present, is a conserved quantity in isolated systems. Everything seems to point to E as defined in Equation (6.13) being the appropriate generalisation of energy to relativistic physics. Figure 6.3 shows a comparison between the Newtonian $mv^2/2$ and our proposed $E = \gamma mc^2$. Ignoring the constant discrepancy introduced by the first term in Equation (6.14), it is again seen that significant divergence occurs for $v > 0.6\,c$. Now what about the extra term appearing in

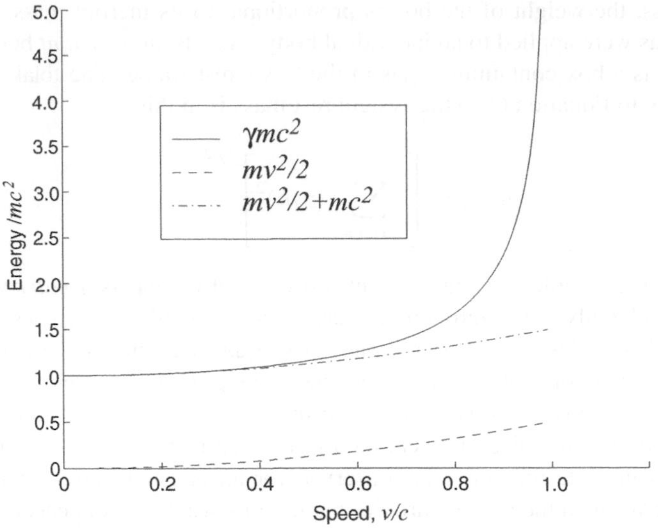

Figure 6.3 Comparison between relativistic and Newtonian expression for energy (cf. Equation (6.14)).

Equation (6.14) which is present even when the particle is at rest? Newtonian physics assigns zero energy to such a particle (provided the zero of potential energy is set at the place where the particle rests), but relativistically it has energy E_{rest}, given by

$$E_{\text{rest}} = mc^2 \,. \tag{6.15}$$

This is the most famous equation in all of physics, adorning many an undergraduate T-shirt. If we are uncomfortable with the notion of free particles at rest having energy then perhaps we could redefine energy as $E' = \gamma mc^2 - mc^2$. But it's too late! Conservation of E_{total} is forced by Equation (6.9) – see the comments just before Equation (6.12) – and this *requires* $E = \gamma mc^2$. Any redefinition at this stage would lead to an 'energy' that is not conserved in an isolated system. In fact the quantity $\gamma mc^2 - mc^2$ does have significance as it is that part of the total energy that is due to the motion of a body in a given frame, so it is appropriate to define this to be the relativistic expression for **kinetic energy**:

- The kinetic energy, K, of a particle of mass m, moving with speed v, is given by

$$K = mc^2(\gamma - 1) = mc^2\left(\frac{1}{\sqrt{1 - v^2/c^2}} - 1\right) . \tag{6.16}$$

But what about the rest energy, $E_{\text{rest}} = mc^2$? What does Equation (6.15) *mean*?

6.2 The Meaning of Rest Energy

Now with the energy defined we can dispense with F and rewrite Equation (6.9), after using Equations (6.10) and (6.13) and rearranging slightly as

$$E_{\text{total}}^2 = p_{\text{total}}^2 c^2 + m_{\text{sys}}^2 c^4 \,. \tag{6.17}$$

For a single particle, m_{sys} is simply the particle's rest mass in which case

$$E^2 = p^2 c^2 + m^2 c^4 \,. \tag{6.18}$$

Equation (6.17) inextricably relates the momentum and energy of a *system* to its rest mass. Despite its apparent triviality, the experimental observation that an isolated box of gas does not change its weight has profound consequences. What if other forms of energy are present in the box? It may contain electromagnetic radiation, or springs compressed by masses, and together all these energy sources will be interacting and exchanging with each other. Whereas in Newtonian physics we had to invent new forms of energy to preserve its conservation, in relativity these extra forms *are already accounted for* once it is established experimentally that the rest mass and momentum of an isolated system are conserved. The conservation of energy has a higher objective status in relativity.

Why doesn't the rest energy matter in Newtonian physics? For a particle moving with speed v we could include a rest energy term and write for a free particle

$$E = mc^2 + \frac{1}{2}mv^2 \,. \tag{6.19}$$

Applying energy conservation to the elastic collision of two particles would then lead to the equation

$$m_1c^2 + \frac{1}{2}mu_1^2 + m_2c^2 + \frac{1}{2}mu_2^2 = m_1c^2 + \frac{1}{2}mv_1^2 + m_2c^2 + \frac{1}{2}mv_2^2 \,, \tag{6.20}$$

which is of course the same as Equation (4.19) since the rest energy terms cancel. The rest terms still cancel in calculating the loss of kinetic energy in inelastic collisions. However, in both cases there is no implication for the total system mass, which is given by $m_1 + m_2$. In Newtonian physics there is no coupling between the loss of energy and the total system mass, and we simply say that the energy is converted to heat. Relativistically, however, the total rest mass of two moving particles is always greater than the individual rest masses. When two particles collide head on and stick together and are then at rest in the lab frame, their total mass *is still* greater than $m_1 + m_2$, the excess accounting for the loss of kinetic energy. All of this works *provided the rest mass of the particles is included in the calculation.*

The argument works in reverse. A stationary massive particle which disintegrates into two smaller particles which fly apart *must liberate some of the energy which held it together to provide kinetic energy to the fragments.* This is the basis of nuclear fission, in which the binding energy of a large nucleus is released as kinetic energy of the fragments after it has disintegrated. *Every* body has the potential to provide energy to fragments into which it *might* disintegrate, and this potentiality *is* the rest energy, $E_{\text{rest}} = mc^2$.

We will see in the next section how this machinery works in practice.

6.3 Relativistic Collisions and Decays

Example 1: *Two particles of mass m collide head on with velocities $\pm v$ in the lab frame and stick together. Calculate the total mass after the collision* (see Figure 6.4).

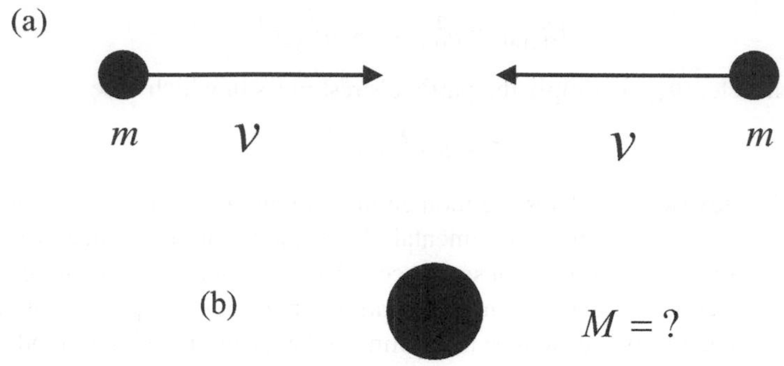

Figure 6.4 Two particles colliding head on: (a) before collision, (b) after collision.

The answer is *not* $2m$. Energy conservation reads

$$E_{\text{tot}}^{\text{before}} = 2\left(\gamma mc^2\right) = E_{\text{tot}}^{\text{after}} = Mc^2\,, \tag{6.21}$$

where M is the mass of the agglomerated body. So

$$M = 2\gamma m = \frac{2m}{\sqrt{1-\frac{v^2}{c^2}}} > 2m\,. \tag{6.22}$$

Example 2: *A neutron decays in the lab into a proton and and an electron. What is the energy of the proton and the electron?*

This example is a little harder because the masses are no longer equal (see Figure 6.5). The neutron is at rest so its energy is equal to its rest energy ($= m_n c^2$) and its momentum is zero.

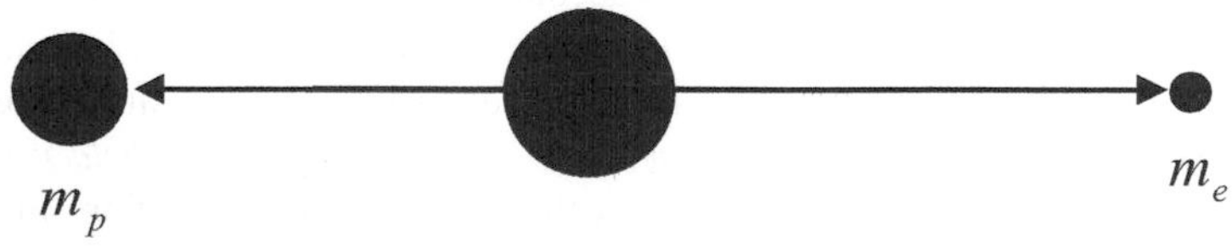

Figure 6.5 Decay of a neutron into a proton and an electron. The unequal arrow lengths indicate that the proton and the electron move at different speeds.

If E_p and E_e are the proton and electron energies then energy conservation reads

$$m_n c^2 = E_p + E_e \quad . \tag{6.23}$$

For the proton and electron momenta, p_p and p_e, we have

$$0 = p_p + p_e \, . \tag{6.24}$$

It seems now that we are stuck because we need to calculate E_p and E_e in terms of the rest masses m_n, m_p and m_e, and all we have are Equations (6.23) and (6.24). However, actually the energy and momentum of a particle are not independent, but related to the particle's rest mass through Equation (6.18), so that effectively Equations (6.23) and (6.24) are two equations for the unknowns E_p and E_e. The algebra, however, is a little tricky so it is worth giving it in detail.

For the proton and electron separately

$$E_p^2 = p_p^2 c^2 + m_p^2 c^4 \, , \tag{6.25}$$

$$E_e^2 = p_e^2 c^2 + m_e^2 c^4 \, . \tag{6.26}$$

Subtracting and using $p_p = -p_e$:

$$E_p^2 - E_e^2 = \left(m_p^2 - m_e^2\right) c^4 \, , \tag{6.27}$$

Dividing through by Equation (6.23):

$$E_p - E_e = \frac{\left(m_p^2 - m_e^2\right) c^2}{m_n} \, , \tag{6.28}$$

and now adding/subtracting this equation from Equation (6.23) we obtain finally

$$E_p = \frac{\left(m_n^2 + m_p^2 - m_e^2\right) c^2}{2m_n} \, , \tag{6.29}$$

$$E_e = \frac{\left(m_n^2 - m_p^2 + m_e^2\right) c^2}{2m_n} \, . \tag{6.30}$$

This type of algebraic manipulation is very typical of energy/momentum calculations in relativity. Notice that the particle velocities were not calculated. If ever a link between energy, momentum and velocity of a particle is required, then it is easy to show from the definitions $E = \gamma m c^2$ and $\mathbf{p} = \gamma m \mathbf{v}$ that

$$\mathbf{p} = \frac{E\mathbf{v}}{c^2} \, , \tag{6.31}$$

which emphasises how momentum transports energy.

6.4 Photons

Photons travel at the speed of light, and so you might think that there is little we can say about their momentum and energy, given that the γ factor, for all inertial frames from which the photon is observed, is infinite. In fact, the last equation just derived indicates that if $v = c$ then

$$E = pc\,, \tag{6.32}$$

showing that for photons energy is carried away at the speed of light. Now Equation (6.18) is still valid even if $E = \gamma mc^2$ and $p = \gamma mv$ no longer apply. The only way Equation (6.18) is compatible with Equation (6.32) is if the rest mass of a photon is zero:

$$m_{\text{photon}} = 0\,. \tag{6.33}$$

Another interesting feature is to calculate the proper time between two nearby events on the world line of a photon. If the events are separated temporally by dt and spatially by $dx = cdt$, then their proper time interval is given by

$$(d\tau)^2 = (cdt)^2 - (dx)^2 = 0\,. \tag{6.34}$$

Photons never grow old!

Although we have figured out quite a lot about photons, we still have not been able to calculate their energy and momentum explicitly. Photons are just light quanta and quantum mechanics provides the expression we need. The energy of a photon of light of wavelength λ is given by

$$E_{\text{photon}} = \frac{hc}{\lambda}\,, \tag{6.35}$$

where h is Planck's constant ($= 6.63 \times 10^{-34}\,\text{J s}$ in SI units). From the point of view of developing relativistic physics, we just regard Equation (6.35) as being 'given', and consequently the photon momentum is

$$p_{\text{photon}} = \frac{h}{\lambda}\,. \tag{6.36}$$

We now have everything we need to calculate interactions between photons and particles, which we now do by way of an example:

A photon of wavelength λ collides with an electron at rest, after which both move in the same straight line. Calculate the wavelength after the interaction (See Figure 6.6).

Energy conservation reads

$$E^{\text{before}}_{\text{photon}} + m_e c^2 = E^{\text{after}}_{\text{photon}} + \sqrt{p_e^2 c^2 + m_e^2 c^4}\,, \tag{6.37}$$

where we have applied Equation (6.18) to the electron energy. Momentum conservation reads

$$\frac{E^{\text{before}}_{\text{photon}}}{c} + 0 = p_e - \frac{E^{\text{after}}_{\text{photon}}}{c}\,. \tag{6.38}$$

where we have used Equation (6.32) relating the photon energy to its momentum. Solving Equations (6.37) and (6.38) for p_e and $E^{\text{after}}_{\text{photon}}$:

$$p_e = \frac{2E^{\text{before}}_{\text{photon}}}{c}\left(\frac{E^{\text{before}}_{\text{photon}} + m_e c^2}{2E^{\text{before}}_{\text{photon}} + m_e c^2}\right), \tag{6.39}$$

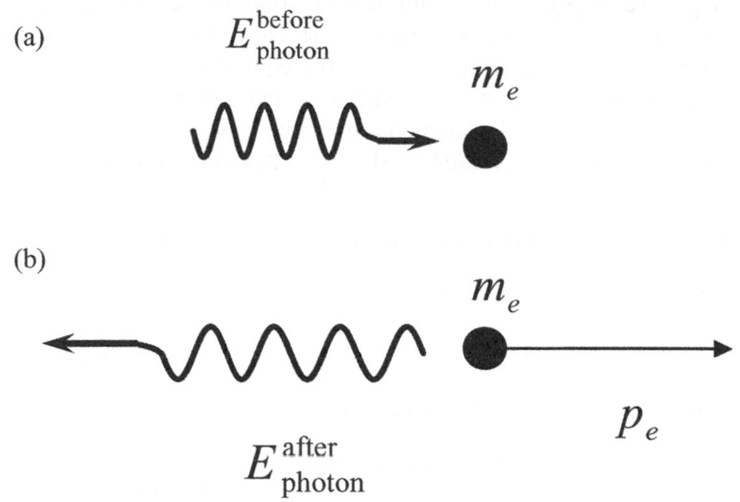

Figure 6.6 A photon scattering one-dimensionally from an electron: (a) before scattering, (b) after scattering. Note the photon's change in wavelength as a result of the interaction.

and

$$E^{\text{after}}_{\text{photon}} = \frac{E^{\text{before}}_{\text{photon}}}{1 + \left(2E^{\text{before}}_{\text{photon}}\big/ mc^2\right)} . \tag{6.40}$$

Equation (6.40) is known as the **Compton scattering formula**. Since $E = hc/\lambda$, it may be written as

$$\lambda_{\text{after}} = \lambda_{\text{before}} \left(1 + \frac{2\lambda_c}{\lambda_{\text{before}}}\right) , \tag{6.41}$$

where $\lambda_c \equiv h/mc$ is the so-called **Compton wavelength**. Equation (6.41) is often expressed in terms of the fractional wavelength shift, or the 'z-parameter'

$$z \equiv \frac{\lambda_{\text{after}} - \lambda_{\text{before}}}{\lambda_{\text{before}}} = \frac{2h}{mc\lambda_{\text{before}}} = 2\frac{\lambda_c}{\lambda_{\text{before}}} . \tag{6.42}$$

Note that $E^{\text{after}}_{\text{photon}} < E^{\text{before}}_{\text{photon}}$, or $\lambda_{\text{after}} > \lambda_{\text{before}}$ and the recoil photon is *red-shifted*.

6.5 Units in High-energy Physics

As we have seen, energy, momentum and mass are absolutely connected in relativity, so much so that high-energy physicists use a system of units that naturally reflects this. First, they use the electron volt (eV) as the base unit, which is just the amount of energy acquired by an electron when accelerated through a potential difference of 1 V. This energy (in joules) is just the electronic charge times 1 V so that

$$1\text{eV} = 1.6 \times 10^{-19}\,\text{J} . \tag{6.43}$$

The usual SI prefixes apply for the multiples of 1000, hence MeV ($= 10^6$ eV), GeV ($= 10^9$ eV) and TeV ($= 10^{12}$ eV) are commonly used. Now, since $E_{\text{rest}} = mc^2$, the rest mass of a particle is conveniently expressed as its energy equivalent divided by c^2. Hence the rest mass of an electron is approximately $0.5\,\text{MeV}\,\text{c}^{-2}$ and a proton is $0.9\ \text{GeV}\,\text{c}^{-2}$. Equation (6.31) indicates that dimensionally momentum can be expressed in energy units divided by c, so a typical momentum is expressed in $\text{MeV}\,\text{c}^{-1}$. A couple of examples will make these ideas clearer:

(i) What is the momentum and velocity of an electron whose total energy is $1\,MeV$ *?*

From Equation (6.17) $pc = (E^2 - (mc^2)^2)^{1/2}$ so

$$pc = \sqrt{(1\,\text{MeV})^2 - (0.5\,\text{MeV})^2} = \sqrt{0.75} = 0.87\,\text{MeV}\,. \tag{6.44}$$

That is $p = 0.87\ \text{MeV}\,\text{c}^{-1}$. From Equation (6.31)

$$v = \frac{pc^2}{E} = \frac{(0.87\ \text{MeVc}^{-1})c^2}{1\ \text{MeV}} = 0.87\ \text{c}\,. \tag{6.45}$$

(ii) Calculate the energy released in the nuclear fission reaction

$$\underbrace{n}_{0.94} + \underbrace{{}^{235}\text{U}}_{219} \rightarrow \underbrace{{}^{140}\text{Xe}}_{130.13} + \underbrace{{}^{94}\text{Sr}}_{87.38} + 2\underbrace{n}_{0.94}\,, \tag{6.46}$$

where the rest masses (in $GeVc^{-2}$*) of each body are given beneath the underbraces.*

The number on the upper left of the symbol of the nuclei is the total number of neutrons and protons.

Total mass before = $(0.94 + 219) = 219.94\text{GeV}\,\text{c}^{-2}$.

Total mass after = $(130.13 + 87.38 + 2 \times 0.94) = 219.39\text{GeV}\,\text{c}^{-2}$.

Difference = $219.94 - 219.39 = 0.55\text{GeV}\,\text{c}^{-2}$.

That is the energy released = 550 MeV.

About 140 MeV of this is the kinetic energy of the reaction products. The rest is in gamma rays (which are just high-energy photons) and neutrinos (a zero-mass particle associated with a category of particles called leptons).

6.6 Energy/Momentum Transformations Between Frames

Momentum and energy are frame-dependent quantities. Evidently in the rest frame of a body the momentum is zero and the energy is just the rest energy ($= mc^2$). However, move to another frame and the momentum is nonzero and the energy now includes kinetic as well as rest energy. Moreover, energy and momentum were derived in Section 6.1 by multiplying through the invariant interval by $m/d\tau$. The spatial part gave us the momentum ($p = mdx/d\tau$), and the temporal part the energy ($E = mdt/d\tau$). If space and time transform according to the Lorentz transformations, then momentum and energy, which are just space and time separations multiplied by an invariant quantity, must also obey Lorentz transformations. We derive the details in this section.

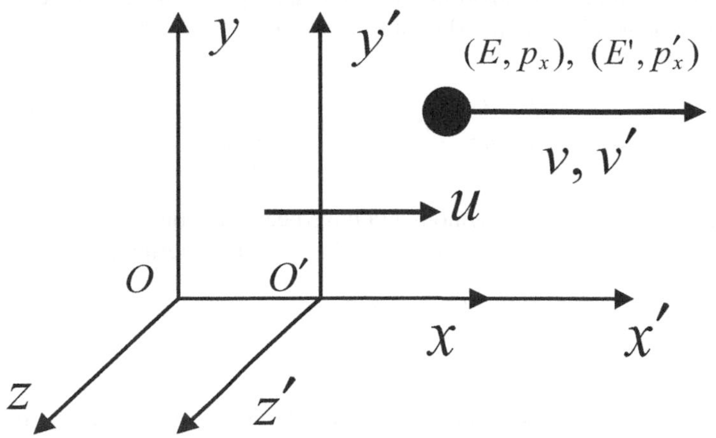

Figure 6.7 A body's energy and momentum as measured by two inertial frames.

In Figure 6.7 a body has velocity $v = dx/dt$ as measured by O, and velocity $v' = dx'/dt'$ as measured by O'. What are its energy and momentum values in O', given their values in O, or, what are E' and p'_x in terms of E and p_x?

In O we have $E = \gamma mc^2$ and $p_x = \gamma mv$ where $\gamma = (1 - v^2/c^2)^{-1/2}$.

In O' we have $E' = \gamma' mc^2$, $p'_x = \gamma' mv$ where $\gamma' = (1 - v'^2/c^2)^{-1/2}$.

Now consider two nearby events on the particle's world line with space–time separation (dt, dx) in O and (dt', dx') in O'. These separations are related through the Lorentz transformations

$$dt' = \gamma_0 \left(dt - udx/c^2\right) \ , \tag{6.47}$$

$$dx' = \gamma_0 \left(dx - udt\right) \ , \tag{6.48}$$

where $\gamma_0 = (1 - u^2/c^2)^{-1/2}$. The invariant proper time, $d\tau$, between the events is

$$d\tau = \left[dt^2 - (dx/c)^2\right]^{1/2} = \left[dt'^2 - (dx'/c)^2\right]^{1/2} \ , \tag{6.49}$$

from which we obtain the dilation factors

$$d\tau = dt\left[1 - (dx/dt)^2/c^2\right]^{1/2} = \gamma^{-1} dt \ , \tag{6.50}$$

and

$$d\tau = dt'\left[1 - (dx'/dt')^2/c^2\right]^{1/2} = (\gamma')^{-1} dt' \ . \tag{6.51}$$

Multiplying the first LT equation (Equation (6.47)) by the invariant quantity $mc^2/d\tau$ and using the dilation factors we obtain

$$mc^2 \frac{dt'}{d\tau} = \underbrace{\gamma' mc^2}_{E'} = \gamma_0 \left(mc^2 \frac{dt}{d\tau} - mu\frac{dx}{d\tau}\right) = \gamma_0 \left(\underbrace{\gamma mc^2}_{E} - u \underbrace{\gamma mv}_{p_x}\right) \ . \tag{6.52}$$

The terms with underbraces are identified with the appropriate energy/momentum. Similarly, multiplying the second LT equation (Equation (6.48)) by the invariant $m/d\tau$

$$m\,\frac{dx'}{d\tau} = \underbrace{\gamma m v'}_{p'_x} = \gamma_0\left(m\frac{dx}{d\tau} - mu\frac{dt}{d\tau}\right) = \gamma_0\left(\underbrace{\gamma m v}_{p_x} - \underbrace{\gamma m c^2}_{E}\,u/c^2\right), \tag{6.53}$$

where again the appropriate energy/momenta have been identified.

Summarising, if E and p_x are known, then E' and p'_x are determined through the relations

$$E' = \gamma_0\,(E - up_x)\,, \tag{6.54}$$

$$p'_x = \gamma_0(p_x - uE/c^2)\,. \tag{6.55}$$

These are the momentum and energy transformations we sought. In deriving these we introduced three different γ factors, two using the particle velocity in the different frames and one using the relative frame velocity. The last one, which we have here labelled γ_0, is the 'usual' γ factor which appeared in the space–time Lorentz transformations (Equations (5.31)) without the subscript.

The inverse transformations are obtained either by direct algebraic inversion of Equations (6.54) and (6.55), or by interchanging primed for unprimed quantities and substituting $-u$ for u. The result is

$$E = \gamma_0\,(E' + up'_x)\,, \tag{6.56}$$

$$p_x = \gamma_0(p'_x + uE'/c^2)\,. \tag{6.57}$$

For completeness we note that multiplying the invariant transverse separations $dy = dy'$ and $dz = dz'$ by the invariant $m/d\tau$ yields invariant transverse momenta

$$p_y = p'_y\,, \tag{6.58}$$

$$p_z = p'_z\,. \tag{6.59}$$

This completes our analysis of energy and momentum transformations.

6.7 Relativistic Doppler Effect

Application of the energy/momentum transformations (Equations (6.54)–(6.59)) to particles is straightforward. As previously noted, energy and momentum change as we move between inertial frames in Newtonian physics, and the same is true of relativistic physics with the appropriate relativistic expressions for momentum and energy. Photons, however, have completely nonclassical expressions for their energy and momentum (Equations (6.35) and (6.36)), and so how they transform between frames will yield new results.[1] The nonclassical expressions relate energy and momentum to the wavelength of the photon, and so calculating

[1] Newton was an incredible genius. Some people have suggested that his stubborn adherence to the 'corpuscular' theory of light, in the face of all the evidence wave theorists (e.g. Christiaan Huygens) could muster, was already a supreme intuition into the first glimmerings of quantum theory. Likewise, his awareness (unlike his contemporaries) of the weakness of some aspects of his theory of mechanics (notably the concepts of absolute space and absolute time) some have suggested presaged relativistic mechanics.

these quantities in a new inertial frame will change the wavelength. This is hardly surprising, as for wave motion in classical physics (e.g. sound) we are used to the idea of the pitch rising as the source approaches – the Doppler effect. The difference in relativity is that there is nothing equivalent to the 'air' supporting the propagation of light. The ether was done away with in Section 5.3. Nevertheless, the wavelength and frequency do change as we move from frame to frame. Our study of relativity, which began with the death knell of the ether through the Michelson–Morley experiment, now concludes with how light is nevertheless affected by relative motion between source and observer: the **relativistic Doppler effect**.

In Figure 6.7, if the moving particle is a photon travelling along the x direction, then according to Equation (6.35) O ascribes to the photon a wavelength λ determined from

$$E = \frac{hc}{\lambda} = p_x c\,, \tag{6.60}$$

with a similar expression relating E', p'_x and λ'. All we have to do now is to substitute these relationships into either Equation (6.54) or Equation (6.55) to calculate the energy of the photon as observed by O':

$$E' = \frac{hc}{\lambda'} = \gamma_0 \left(E - up_x\right) = \gamma_0 \left(\frac{hc}{\lambda} - u\frac{h}{\lambda}\right)\,, \tag{6.61}$$

or

$$\frac{1}{\lambda'} = \left(\frac{1-u/c}{\sqrt{1-u^2/c^2}}\right)\frac{1}{\lambda} = \left(\frac{1-u/c}{1+u/c}\right)^{1/2}\frac{1}{\lambda}\,. \tag{6.62}$$

Finally, the wavelength of the photon as observed from O' is given by

$$\lambda' = \left(\frac{c+u}{c-u}\right)^{1/2}\lambda\,. \tag{6.63}$$

If $u \ll c$, as is often the case when analysing radiation from moving atoms, then to first order in (u/c) the above formula reduces to (exercise!)

$$\lambda' \approx \lambda\left(1 + \frac{u}{c}\right)\,, \tag{6.64}$$

or, in terms of the red-shift parameter, z, introduced earlier

$$z = \frac{\lambda' - \lambda}{\lambda} = \frac{u}{c}\,. \tag{6.65}$$

The wavelength in O' is longer than in O. Sometimes in solving problems concerning the Doppler effect it is necessary to think carefully about (a) how to establish the standard orientation, (b) in which direction the photon is moving relative to the standard orientation and (c) in which frame the wavelength is known.

6.8 Summary

Section 6.1. Multiplying the invariant proper time $d\tau = [(dt)^2 - (dx/c)^2]^{1/2}$ between two neighbouring events on the world line of a body moving along the x-axis, by the invariant quantity $m/d\tau$, yields two further quantities

$$\mathbf{p} = \gamma m \mathbf{v}\,, \tag{6.7}$$

$$E = \gamma mc^2 \,, \tag{6.13}$$

respectively known as the **momentum** and **energy** of the body. Thus $\mathbf{p} = E\mathbf{v}/c^2$ (Equation (6.31)). For an isolated system both energy and momentum are conserved. The total system rest mass is

$$m_{\text{sys}} = \left[E_{\text{total}}^2 - \left(\frac{p_{\text{total}}}{c}\right)\right]^{1/2} \,, \tag{6.9}$$

where $E_{\text{total}} = \sum_{\text{particles}} \gamma mc^2$ (Equations (6.10) and (6.13)) and $p_{\text{total}} = \sum_{\text{particles}} \gamma mv$.

The **rest energy** of a body is given by

$$E_{\text{rest}} = mc^2 \quad . \tag{6.15}$$

The **kinetic energy** of a body is given by

$$K = mc^2(\gamma - 1) = mc^2\left(\frac{1}{\sqrt{1 - v^2/c^2}} - 1\right) \,. \tag{6.16}$$

Section 6.2. For a single particle Equation (6.9) becomes

$$E^2 = p^2c^2 + m^2c^4 \,. \tag{6.17}$$

Section 6.4. For photons $m = 0$, so that

$$E = pc \,. \tag{6.32}$$

From quantum mechanics, photons have an energy

$$E_{\text{photon}} = \frac{hc}{\lambda} \,. \tag{6.35}$$

A photon, scattering from a stationary electron in one dimension, changes its wavelength according to the Compton scattering formula:

$$z \equiv \frac{\lambda_{\text{after}} - \lambda_{\text{before}}}{\lambda_{\text{before}}} = \frac{2h}{mc\lambda_{\text{before}}} = 2\frac{\lambda_c}{\lambda_{\text{before}}} \,. \tag{6.42}$$

where $\lambda_c \equiv h/mc$ is the Compton wavelength.

Section 6.6. For inertial frames oriented in the standard way, energy and momentum transform according to

$$E' = \gamma_0 \left(E - up_x\right) \,, \tag{6.54}$$

$$p'_x = \gamma_0(p_x - uE/c^2) \,, \tag{6.55}$$

the inverse transformations being obtained by interchanging primed for unprimed quantities in Equations (6.54) and (6.55), and substituting $-u$ for u.

Section 6.7. Using Equation (6.35) the relationship between the wavelengths of a photon viewed from different inertial frames is

$$\lambda' = \left(\frac{c+u}{c-u}\right)^{1/2} \lambda. \tag{6.63}$$

If $u \ll c$ then the fractional change in wavelength is

$$z \equiv \frac{\lambda' - \lambda}{\lambda} = \frac{u}{c} . \tag{6.65}$$

6.9 Problems

6.1 Express the following quantities in practical units, for example $\mathrm{MeV\,c^{-2}}$. ($c = 3 \times 10^8\,\mathrm{m\,s^{-1}}$, $e = 1.6 \times 10^{-19}\,\mathrm{C}$).

(a) Electron mass: $m_e = 9.1 \times 10^{-31}\,\mathrm{kg}$.
(b) Proton mass: $m_p = 1.67 \times 10^{-27}\,\mathrm{kg}$.
(c) Total energy of an electron with momentum $p = 1\mathrm{MeV\,c^{-1}}$.
(d) Kinetic energy of a proton with momentum $p = 1\mathrm{MeV\,c^{-1}}$.
(e) Repeat (d) using the Newtonian formula $K = p^2/2m$.
(f) Kinetic energy of a proton with speed $v = 0.5c$.

6.2 Express the following in SI units (n.b. $1\,\mathrm{eV} = 1.6 \times 10^{-19}$ J).

(a) The rest energy of an electron, $m_e = 0.511\,\mathrm{MeV\,c^{-2}}$.
(b) The energy liberated when an electron and a positron annihilate at rest.
(c) The energy of cosmic rays of the highest observed energy ($\sim 10^{19}\,\mathrm{eV}$).

6.3 The neutral pion has a mass $m_\pi = 135$ $\mathrm{MeV\,c^{-2}}$ and its mean lifetime in its rest frame is $\tau_\pi = 10^{-16}\,\mathrm{s}$. Calculate the mean flight path (in the laboratory) before decay of:

(a) a pion with velocity $0.1c$.
(b) a pion with momentum 1 $\mathrm{GeV c^{-1}}$.
(c) a pion with kinetic energy 1 TeV.

6.4 (a) If the fission of one nucleus of $^{235}_{92}\mathrm{U}$ releases 200 MeV, estimate the consumption in kg per year of this fuel in a fission reactor with a power output of 1 GW. Assume the efficiency is 33% and take a value of Avagadro's number of $N_A = 6 \times 10^{26}\,\mathrm{kg^{-1}\ mole^{-1}}$.
(b) One D–D (D is the symbol for deuterium, the heavy isotope of hydrogen $^2\mathrm{H}$) fusion reaction releases 4 MeV. The mass ratio of deuterium to hydrogen in sea water is 1/3250. What is the fusion energy available in sea water in $\mathrm{J\,kg^{-1}}$?

6.5 An observer O' moves along the positive x-axis with velocity $u = 0.9c$ relative to an observer O. The space–time origins of O and O' coincide. Use Lorentz transformations to answer the following.

(a) What are the space–time coordinates in O' of events A and B occurring with coordinates $(x_A, t_A) = (0, 100\text{ ns})$ and $(x_B, t_B) = (-100\text{ m},\ 0)$ in O?
(b) What are the space–time coordinates in O of events C and D occurring with coordinates $(x'_C, t'_C) = (0,\ 100\text{ ns})$ and $(x'_D, t'_D) = (-100\text{m},\ 0)$ in O'?
(c) What are the momentum and energy in O' of an electron ($m = 0.5\,\text{MeV}\text{c}^{-2}$) with momentum $p = 0.5\text{MeVc}^{-1}$ along the positive x direction in O?
(d) What are the momentum and energy in O of an electron ($m = 0.5\,\text{MeVc}^{-2}$) with kinetic energy, $K = 1\,\text{MeV}$ in O'?

6.6 (a) A π^0 decays at rest into two photons. If $m_{\pi^0} = 135$ MeV/c^2, calculate the energy and momentum of each photon.
(b) A K^0 decays at rest into two π^0s. If $m_{K^0} = 498$ MeVc^{-2}, calculate the energy and momentum of each π^0.

6.7 Two particles of mass $m = 0.5$ MeV c^{-2} collide head-on with speed in the laboratory frame of $0.5c$ and then stick together. Calculate the mass of the agglomerated particle.

6.8 A photon moving from left to right strikes an electron at rest, causing the electron to also move from left to right. Prove that the scattered photon cannot be moving from left to right.

6.9 A uranium nucleus emits an alpha particle of mass 6.6×10^{-27} kg with speed 1.5×10^7 m s^{-1}. For this question use $1\,\text{eV} = 1.6021 \times 10^{-19}\,\text{J}$ and $c = 2.9979 \times 10^8\,\text{m s}^{-1}$.

(a) Calculate the mass of the He nucleus in GeVc^{-2}.
(b) Calculate accurately (i.e. without using a Newtonian approximation, and giving your answer to four decimal places) the total energy (in GeV) and the momentum (in GeVc^{-1}) of the alpha particle.
(c) Now using Newtonian formulae calculate the *kinetic* energy and the momentum, expressing your answers in each case as fractions of the corresponding relativistic formulae (i.e. find A and B in $K_{\text{Newtonian}} = A \times K_{\text{Relativistic}}$ and $P_{\text{Newtonian}} = B \times P_{\text{Relativistic}}$.
(d) What fraction of the alpha particle's rest energy is present as kinetic energy after the decay?

6.10 The Sun produces energy by fusion of hydrogen into helium. One series of processes is:

$$\text{p} + \text{p} \rightarrow {}^2\text{H} + \text{e}^+ + \nu + 0.41\text{ MeV}$$

$${}^2\text{H} + \text{p} \rightarrow {}^3\text{He} + \gamma + 5.51\text{ MeV}$$

$${}^3\text{He} + {}^3\text{He} \rightarrow {}^4\text{He} + 2\text{p} + \gamma + 12.98\text{ MeV}\,.$$

Calculate the total energy released in the formation of one α particle (^{4}He nucleus). Note that the e$^+$ (positrons) will annihilate with electrons in the sun giving extra energy. (Hint: how many times must each reaction occur in order to produce a single helium nucleus?)

6.11 A gamma ray of energy 1.2 MeV scatters one-dimensionally from an electron. Evaluate the red-shift parameter $z \equiv (\lambda' - \lambda)/\lambda$ for this process.

6.12 Following a relativity lecture, with her brain occupied by new ideas in physics, a first-year student drives along the road in her car and goes through a red light. At the court case, the student, in her defence, tells the judge that in driving towards the light it appeared green due to the Doppler effect. The judge accepted this plea and changed the charge to one of speeding. Taking values of wavelengths ($\lambda_{\text{red}} = 650$ nm; $\lambda_{\text{green}} = 530$ nm), use the Doppler formula Equation (6.63) where the source (at rest in O) *approaches* the observer O', to estimate the fine imposed if the standard rate is £10 for each mph (1 mile per hour $= 0.45\text{ms}^{-1}$) in excess of the limit (30 mph).

7

Gravitational Orbits

7.1 Introduction

Recall from Chapter 2 (Equation (2.1)) Newton's second law applied in three dimensions to a body of constant mass

$$\mathbf{F} = m\mathbf{a}, \quad \begin{aligned} F_x(x, y, z, v_x, v_y, v_z, t) &= m\ddot{x} \\ F_y(x, y, z, v_x, v_y, v_z, t) &= m\ddot{y} \\ F_z(x, y, z, v_x, v_y, v_z, t) &= m\ddot{z} \ . \end{aligned} \tag{7.1}$$

As we emphasised previously, solving this general problem represents a considerable mathematical challenge. Simplifying to one dimension in Chapter 2 enabled us to solve many important mechanical problems, principally by introducing useful scalar quantities such as work and energy. Now we want to step into the real world of three dimensions, seeing how these scalar quantities are generalised, and to introduce *vector* quantities that are particular to higher dimensional motion (torque and angular momentum).

The mathematical problem is expressed in its entirety by the differential equations of Equation (7.1). The ultimate function of the scalar and vector quantities that we will introduce in this chapter is to yield insight into the solutions of Equation (7.1) with the minimum of mathematical effort. The example par excellence that we shall consider is the gravitational orbit of a planet around the Sun.

Since we are now working in three dimensions, vector notation will be vital and, in particular, the vector products $\mathbf{a} \cdot \mathbf{b}$ and $\mathbf{a} \times \mathbf{b}$ will occur frequently. The Appendix (Sections A.1 and A.2) gives a brief review of these products if you need a refresher.

7.2 Work in Three Dimensions

Recall that in one dimension the increment of work, dW, done by a force, F, in moving a particle a distance dx is given by $dW = F dx$ and that the work–energy theorem states that the change in kinetic energy is equal to the work done by the force. What are the equivalent statements for a force, $\mathbf{F}$, acting on a particle free to move in three dimensions (see Figure 7.1)? The figure suggests that in three dimensions it is possible for the displacement of the particle, $d\mathbf{l}$, to be in a different direction to the applied force. This may seem a little strange at first, but you only need to think of throwing a ball to see that it is quite logical. An infinitesimal displacement along the particle's parabolic trajectory is in general not in the same direction as the force of gravity, which is always vertically down. It is, however, clear

Classical Mechanics – Second Edition From Newton to Einstein: A Modern Introduction Martin W. McCall

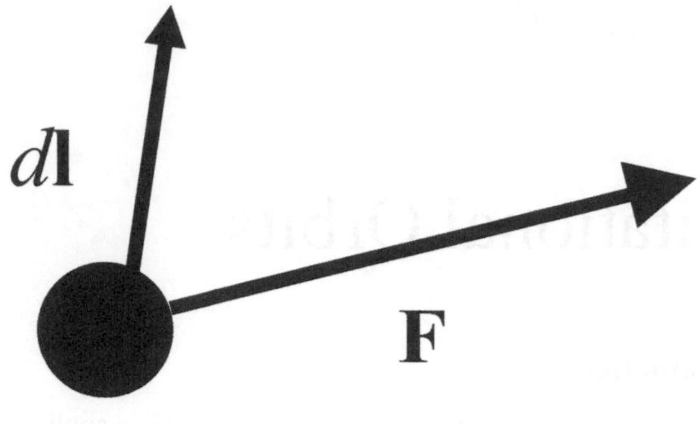

Figure 7.1 A force acting on a body in three dimensions.

that only the *component* of **F** along $d\mathbf{l}$ actually contributes to any change in the particle's kinetic energy, so that the work done by **F** in displacing the particle an amount $d\mathbf{l}$ is given by

$$dW = \mathbf{F} \cdot d\mathbf{l} \,, \tag{7.2}$$

or in Cartesian coordinates

$$dW = F_x dx + F_y dy + F_z dz \,. \tag{7.3}$$

So the total work in moving the particle from A to B along a particular path is given by

$$W = \int_{\text{path}} dW = \int_{\text{path}} \mathbf{F} \cdot d\mathbf{l} \,. \tag{7.4}$$

This is a type of integral you may not have met before, called a **line integral**. We will not be using the full machinery of line integrals, relying instead on geometrical arguments. However, the flavour of what is being done is illustrated in Figure 7.2. The particle moves from point A to point B along the path illustrated. At any point along the path we can identify $d\mathbf{l}$ as an infinitesimal displacement tangent to the path at that point. Taking the scalar product with **F**, the force acting at that point, we obtain the scalar infinitesimal, dW. These are then summed up (i.e. integrated) to obtain the total work, W, in going from A to B. The force may depend on the position along the path. The path may either be the actual physical path followed by the particle, or a hypothetical one introduced to answer the question 'What would be the work done by **F** if the path taken were so-and-so?' The important point is that if, keeping A and B fixed, we vary the path, then the value of W obtained *may also vary*, in which case we say that W is **path-dependent**. Are there any situations where W is **path-*in*dependent**?

7.3 Torque and Angular Momentum

Following the development of Chapter 2, we will restrict ourselves to forces that are independent of the velocity and have no explicit time-dependence. Were we to include the

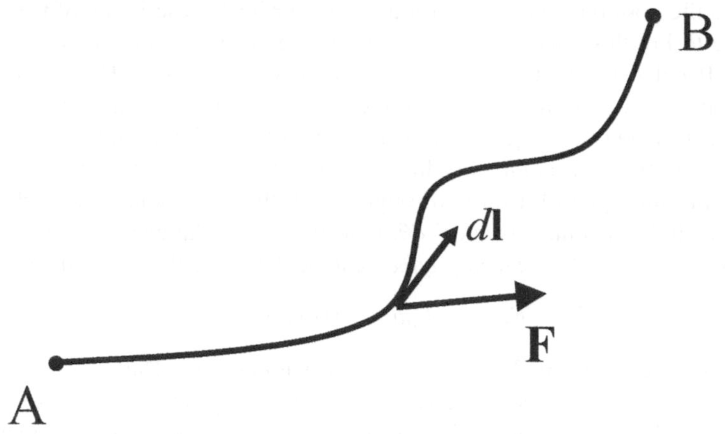

Figure 7.2 A force moving a particle from A to B.

latter we would have to address such questions as 'How fast did the particle go from A to B?, or 'When did the particle go from A to B?' and this would involve us in further complications. Regarding a position-dependent force, $\mathbf{F}(\mathbf{r})$, how might we describe its action on a particle at position $\mathbf{r}$? Figure 7.3 shows one possibility. In this case the force acts in the direction

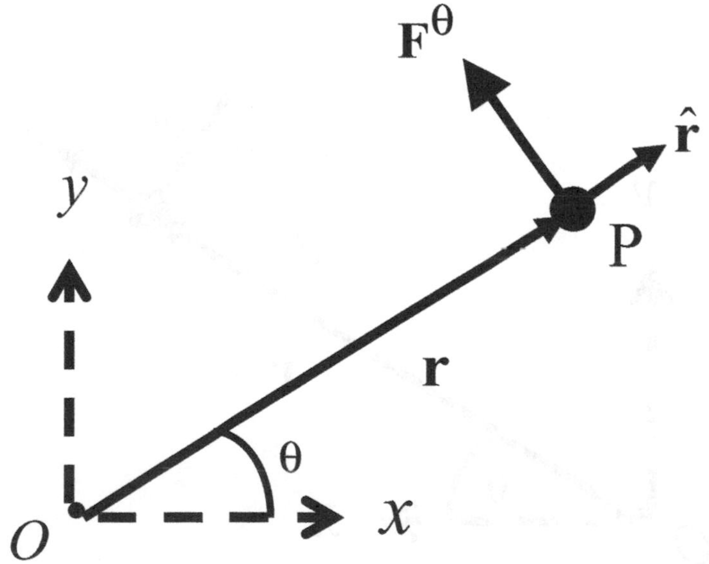

Figure 7.3 Tangential force.

orthogonal to the position vector $\mathbf{r}$, or, for a polar coordinate description of $\mathbf{r}$, in the direction of increasing θ. For this reason, the force can be regarded as being *tangential*, and may be denoted $\mathbf{F}^\theta$. If $\hat{\mathbf{r}}$ denotes the unit vector in the radial direction, then $\mathbf{F}^\theta \cdot \hat{\mathbf{r}} = 0$. In this polar representation it appears rather as if the force $\mathbf{F}^\theta$ is turning the particle about the origin. In fact this idea turns out to be quite useful in terms of calculating the work done by $\mathbf{F}^\theta$ in increasing the particle's *angular* coordinate from θ to $\theta + d\theta$. This suggests the existence of some angular analogue of a force through which the work done can be characterised in conjunction with the angular 'distance' $d\theta$. Indeed, this angular analogue of the force, which we shall provisionally call the *turning effect*, can be determined through the relation

$$dW = F_x dx + F_y dy = (\text{turning effect}) \times d\theta \qquad (7.5)$$

where we have used Equation (7.3) restricted to two dimensions. There is no loss of generality in restricting to two dimensions, as the vectors $\mathbf{F}^\theta$ and $\mathbf{r}$ only define a plane, so we can always choose the x- and y- axes to lie in this plane. Now if the induced displacement $|d\mathbf{l}| = rd\theta$ is parallel to $\mathbf{F}^\theta$ (it need not be, but we imagine it is for the moment), then with the aid of Figure 7.4 the displacement increments dx and dy are found to be

$$dx = -rd\theta \times \sin\theta = -y\theta\ ,$$

and

$$dy = rd\theta \times \cos\theta = xd\theta\ ,$$

so that

$$dW = (-F_x y + F_y x)d\theta\ . \qquad (7.6)$$

The turning effect induced by the tangential force is then characterised by the quantity $(xF_y - yF_x)$, which looks like the z component of the vector $\mathbf{r} \times \mathbf{F}^\theta$. Let this vector be

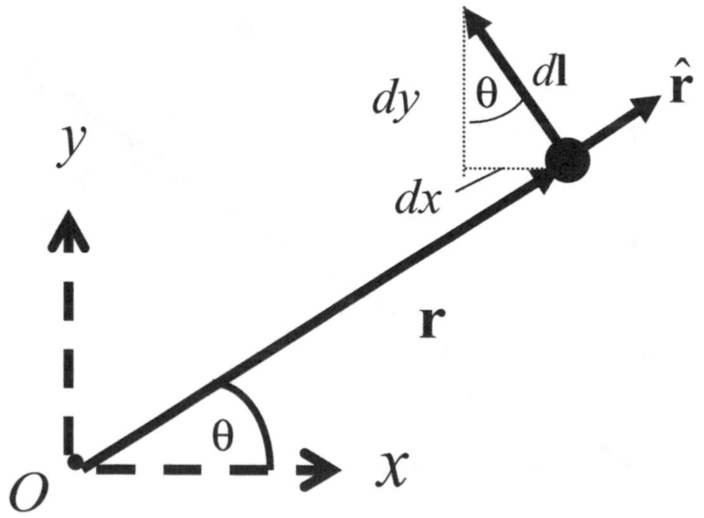

Figure 7.4 Displacement induced by the tangential force of Figure 7.3.

denoted $\mathbf{G}$ and call it the **torque**:

$$\mathbf{G} = \mathbf{r} \times \mathbf{F} = \mathbf{r} \times \dot{\mathbf{p}} \,, \tag{7.7}$$

the second equality resulting from using Newton's second law. Dropping the superscript on $\mathbf{F}$ is not a mistake, since any radial component to a general force, $\mathbf{F}$, does not contribute to the cross product defining $\mathbf{G}$. There is a little trick here to take the time derivative on $\mathbf{p}$ outside the cross product. The usual rule for differentiating a product also works for a cross product (check it by writing it out in components if you like) so that

$$\frac{d}{dt}(\mathbf{r} \times \mathbf{p}) = \dot{\mathbf{r}} \times \mathbf{p} + \mathbf{r} \times \dot{\mathbf{p}} \,.$$

Now the first term on the right is zero, because $\dot{\mathbf{r}}$ is parallel to $\mathbf{p} = m\dot{\mathbf{r}}$, so substituting back into Equation (7.7) we obtain

$$\mathbf{G} = \frac{d}{dt}(\mathbf{r} \times \mathbf{p}) \equiv \frac{d}{dt}\mathbf{L} \,, \tag{7.8}$$

where $\mathbf{L} \equiv \mathbf{r} \times \mathbf{p}$ has been defined. Equation (7.8) says that the angular analogue of force, or **torque**, $\mathbf{G}$, is the rate of change of another vector, $\mathbf{L}$. The equation becomes an angular analogue of Newton's second law if $\mathbf{L}$ is interpreted as the angular analogue of momentum, **the angular momentum**. Equation (7.8) is so important that it is worth restating in words:

Torque equals rate of change of angular momentum.

We have not derived a new law of mechanics, independent of Newton's laws. It is really an efficient way of collecting the information content of Newton's second law, Equation (7.1), when applied to systems in three dimensions. We now consider how it works when $\mathbf{F}$ is purely radial.

7.4 Central Forces

Most position-dependent forces originate from some source point so that the force acts either radially outwards or inwards, and such forces are known as **central forces**. It is natural to choose the origin of coordinates to coincide with the source point, and with the origin located in this way, the general form for a central force is

$$\mathbf{F} = F(r)\hat{\mathbf{r}} \,, \tag{7.9}$$

where $r \equiv |\mathbf{r}|$ (see Figure 7.5). The sign of $F(r)$ is positive if the force acts in the direction of increasing r, that is *repelling* a body away from the source, and is negative if it *attracts* a body towards the source point. Examples of central forces are

$$\text{Gravitation} \qquad F(r) = -\frac{GmM}{r^2} \,, \qquad G = 6.7 \times 10^{-11}\,\mathrm{Nm^2\,kg^{-2}} \tag{7.10}$$

$$\text{Coulomb} \qquad F(r) = \frac{Q_1 Q_2}{4\pi\varepsilon_0 r^2} \,, \qquad \varepsilon_0 = 8.9 \times 10^{-12}\,\mathrm{F\,m^{-1}} \tag{7.11}$$

$$\text{Spring in 3D} \qquad F(r) = -kr \,, \tag{7.12}$$

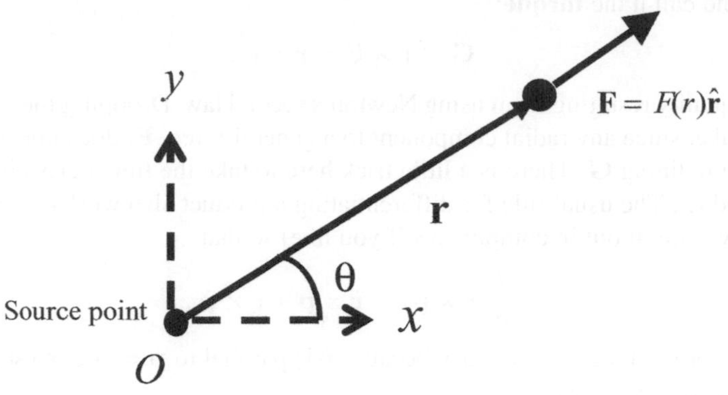

Figure 7.5 Radial force acting on a particle.

The electrostatic Coulomb force is attractive if the charges Q_1 and Q_2 are different, or repulsive if they are the same. Gravity is always attractive (except recent observations suggest that at cosmic scales a repulsive term may be significant). By setting Q_1 and Q_2 equal to the electronic charge (=1.6 × 10^{-19} coulombs), $m = M = m_e$, the mass of the electron (=9.1 × 10^{-31} kg), and dividing Equation (7.10) by (7.11), the ratio of the electrostatic to gravitational forces is found to be 4.2 × 10^{42}; gravity is very weak compared with electricity.

A central force exerts no torque as can be seen from

$$\mathbf{r} \times \mathbf{F} = \mathbf{r} \times F(r)\hat{\mathbf{r}} = 0, \tag{7.13}$$

and consequently, from Equation (7.8) we conclude that

$$\mathbf{L} = \mathbf{r} \times \mathbf{p} = \mathbf{constant}\ . \tag{7.14}$$

For central forces the angular momentum is conserved. It is important to appreciate that this is an origin-dependent conclusion. A choice of origin distinct from the source of the force would result in a nonzero torque and consequently a time-varying angular momentum. This emphasises the fact that the angular momentum equation (Equation (7.14)) is a convenient way to marshal the facts contained within Newton's second law, Equation (7.1). Taking gravity as an example, in Cartesian coordinates, Equations (7.1) turn out to be,

$$-\frac{GmMx}{(x^2+y^2+z^2)^{3/2}} = m\ddot{x},$$
$$-\frac{GmMy}{(x^2+y^2+z^2)^{3/2}} = m\ddot{y},$$
$$-\frac{GmMz}{(x^2+y^2+z^2)^{3/2}} = m\ddot{z}.$$

From these equations, it is hardly 'obvious' that the quantities

$$(y\dot{z} - z\dot{y})\ ,\ \ (z\dot{x} - x\dot{z}) \text{ and } (x\dot{y} - y\dot{x})$$

are conserved. However, these are just the Cartesian components of the angular momentum. Another distinction between Equation (7.14) and its linear counterpart is that whereas the conservation of linear momentum for isolated systems is a consequence of the absence of external forces, the conservation of angular momentum occurs even when forces *are* present. However, for angular momentum conservation, the forces present must be of a particular geometry, that is central forces. The conservation of angular momentum for central forces is the key result through which the complicated equations of motion in three dimensions can be enormously simplified.

We are now able to answer the question posed at the end of Section 7.2. In short, the work done by a position-dependent force, $F(r)$, in taking a particle from point A to point B, is independent of the path taken if and only if the force is central, that is $\mathbf{F}(\mathbf{r}) = F(r)\hat{\mathbf{r}}$. To prove this algebraically requires some vector calculus, so instead we will use a perfectly rigorous geometrical argument (see Figure 7.6). The figure shows two paths between points

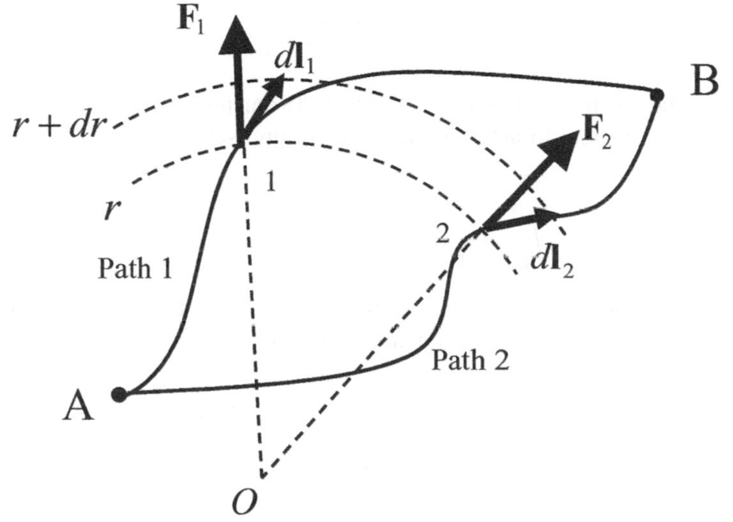

Figure 7.6 Path independence of work for central forces: geometric proof.

A and B, and a central force emanating from the origin O. Two radii of length r and $r + dr$ intersect the paths at points 1 and 2 at which points infinitesimal displacement vectors, $d\mathbf{l}_1$ and $d\mathbf{l}_2$ are drawn tangent to the paths, their tips being both located on the circle of radius $r + dr$. The forces at points 1 and 2, denoted by $\mathbf{F}_1$ and $\mathbf{F}_2$, are equal in magnitude (since they are at the same radius), but act in different radial directions. Now from the geometry of the diagram, the infinitesimal work increments, dW_1 and dW_2, are given by

$$dW_1 = \mathbf{F}_1 \cdot d\mathbf{l}_1 = F dr = \mathbf{F}_2 \cdot d\mathbf{l}_2 = dW_2\,, \tag{7.15}$$

where the middle equalities result from the fact that the radial projections of $d\mathbf{l}_1$ and $d\mathbf{l}_2$ are both $dr\hat{\mathbf{r}}$. All infinitesimal displacements along the paths can be paired off in this way by choosing a different value of r, so that integrating r between its value at point A and point B we obtain the total work as

$$W_{\rm AB} = \int_{\rm path\ 1} \mathbf{F}_1 \cdot d\mathbf{l}_1 = \int_{\rm path\ 2} \mathbf{F}_2 \cdot d\mathbf{l}_2 = \int_{r_{\rm A}}^{r_{\rm B}} F dr \tag{7.16}$$

which is the three-dimensional analogue of $W = \int_{x_{\rm A}}^{x_{\rm B}} F(x)dx$. The work done is independent of the path taken, but does depend on the location of the end-points A and B. Formally, then, we can evaluate the integral in Equation (7.16) as

$$W_{\rm AB} = \int_{r_{\rm A}}^{r_{\rm B}} F dr = -U(r_{\rm B}) + U(r_{\rm A}) \,, \tag{7.17}$$

where U is a function of r only, and the minus sign is for convenience. Hopefully you can see the way we are edging. The function $U(r)$ plays exactly the same role for central forces as the potential function, $U(x)$ played for conservative forces in one dimension. In fact

$$F = -\frac{dU(r)}{dr} \,. \tag{7.18}$$

Is there an analogue for the work–energy theorem? (Remember, the answer is always 'yes' if the lecturer/author bothers to ask the question!) Using Newton's second law in Equation (7.16)

$$W_{\rm AB} = \int_{\rm A}^{\rm B} \mathbf{F} \cdot d\mathbf{l} = \int_{\rm A}^{\rm B} m\frac{d\mathbf{v}}{dt} \cdot d\mathbf{l} = \int_{\rm A}^{\rm B} m\frac{d\mathbf{l}}{dt} \cdot d\mathbf{v} = \int_{\rm A}^{\rm B} m\mathbf{v} \cdot d\mathbf{v} \,. \tag{7.19}$$

The manipulation for the last equality is exactly the same as in Equation (2.8), which works for vectors as well, once we note that the velocity $\mathbf{v} = d\mathbf{l}/dt$. Now

$$\mathbf{v} \cdot d\mathbf{v} = v_x dv_x + v_y dv_y + v_z dv_z = \frac{1}{2} d(v_x^2 + v_y^2 + v_z^2) = d(v^2)$$

so the last integral in Equation (7.19) is evaluated as

$$W_{\rm AB} = \int_{\rm A}^{\rm B} m\mathbf{v} \cdot d\mathbf{v} = \frac{1}{2}mv_{\rm B}^2 - \frac{1}{2}mv_{\rm A}^2 \,, \tag{7.20}$$

which is just the change in the particle's kinetic energy in being moved from A to B. Combining this with Equation (7.17) we obtain the three-dimensional form of the work–energy theorem:

$$W_{\rm AB} = U(r_{\rm A}) - U(r_{\rm B}) = \frac{1}{2}mv_{\rm B}^2 - \frac{1}{2}mv_{\rm A}^2 \,. \tag{7.21}$$

In words, the work done by a central force in moving a particle from point A to point B is equal to the change in the particle's kinetic energy. Using exactly the same reasoning as for Section 2.6, a rearrangement of the second equality in Equation (7.21) gives

$$\frac{1}{2}mv_{\rm A}^2 + U(r_{\rm A}) = \frac{1}{2}mv_{\rm B}^2 + U(r_{\rm B}) \equiv E \,, \tag{7.22}$$

where the total mechanical energy, E, defined as the sum of kinetic and potential energies, is conserved for a central force. That this is so is evident from Equation (7.22), which equates

E for the initial and final states. In every way we can conclude that central forces are the three-dimensional generalisation of the conservative forces of one-dimension which are, in fact, a special case of the former.

A note on the use of vector calculus notation is in order. We have carefully avoided this up until now, to make the development as simple as possible. However, just so that you can recognise the statements we have made in the mathematically rigorous form in which you will find them in more advanced mechanics textbooks, we will quote the results, without any exposition.

For a central force, $\mathbf{F} = F(r)\hat{\mathbf{r}}$, it is possible to introduce a potential function, $U(r)$, such that

$$F = -\nabla U \ , \text{ or } \ U = -\int \mathbf{F} \cdot d\mathbf{l} \,. \tag{7.23}$$

Here ∇ is the differential vector operator 'grad', defined in Cartesian coordinates as

$$\nabla \equiv \mathbf{i}\frac{\partial}{\partial x} + \mathbf{j}\frac{\partial}{\partial y} + \mathbf{k}\frac{\partial}{\partial z} .$$

The statement that the line integral $\int_{\mathrm{A}}^{\mathrm{B}} \mathbf{F} \cdot d\mathbf{l}$ is the same for path 1 and path 2 is the same as saying that integrating from A to B along path 1, and back from B to A along path 2, gives zero overall. This generalises to *any* closed path so that

$$\oint_{\text{Closed path}} \mathbf{F} \cdot d\mathbf{l} = 0 \,. \tag{7.24}$$

The equivalent vector differential statement is

$$\nabla \times \mathbf{F} = \mathbf{0} \,, \tag{7.25}$$

where the vector differential operator $\nabla\times$ is known as the 'curl'. These mathematical statements contain the full content of the theory developed in this section.

7.5 Gravitational Orbits

We have assembled a formidable tool-kit for analysing motion in three dimensions showing, in particular, that the total mechanical energy, E, *and* the vector angular momentum, $\mathbf{L}$, are conserved for central forces. We are now going to apply these ideas to the most important case of the motion of a planet under the Sun's gravitational influence. By regarding the Sun's mass, M, as being very much greater than the planet's mass, m, any reciprocal gravitational action of the planet on the Sun may be ignored. The Sun will be the fixed source-point of the central force, sitting at the coordinate origin, as in Figure 7.7. Emphasising that we do not know the details yet, Figure 7.8 depicts an orbit fragment as a wiggly line. Applying the work–energy theorem as the mass moves from position A to position B:

$$\int_{\mathrm{A}}^{\mathrm{B}} \mathbf{F} \cdot d\mathbf{l} = \mathrm{K_B} - \mathrm{K_A} = \frac{1}{2}mv_{\mathrm{B}}^2 - \frac{1}{2}mv_{\mathrm{A}}^2 \,. \tag{7.26}$$

Since the force is central, the left-hand side is independent of the path chosen, so even though the *actual* path is presumed to be the one shown by the solid curve in Figure 7.8, we are at

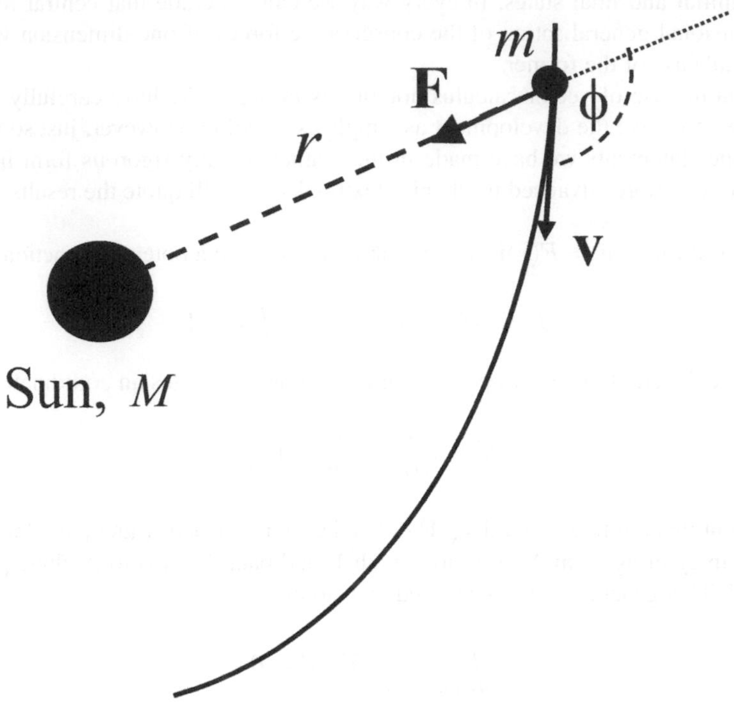

Figure 7.7 A planet orbiting the Sun.

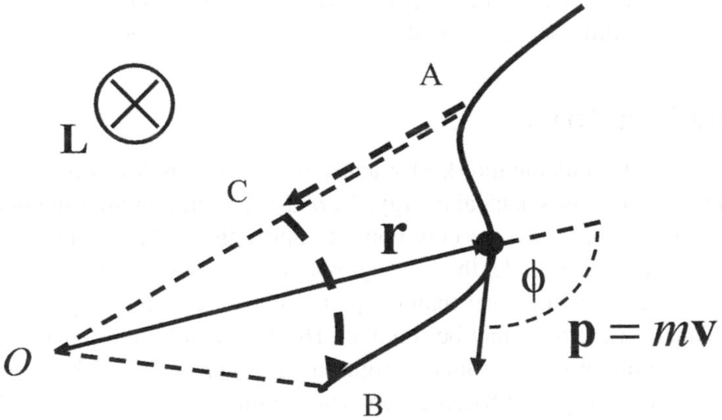

Figure 7.8 A fragment of the planet's orbit (solid curve). The path indicated with the broken arrows A–C–B is an auxiliary path to assist in the calculation of the work done by gravity.

liberty to choose *any* path between A and B which makes the integration convenient. We select the auxiliary path indicated by the dotted arrows, that is a straight radial line from A to C, followed by an arc from C to B, of radius r_B. For this *virtual* path, the line integral becomes .

$$\int_A^B \mathbf{F} \cdot d\mathbf{l} = \int_{r_A}^{r_B} \left(-\frac{GMm}{r^2}\right) dr + \int_{\substack{\text{Along} \\ \text{circumf.} \\ \text{radius } r_B}} \mathbf{F} \cdot d\mathbf{l} \,. \tag{7.27}$$

The second term on the right is zero since $\mathbf{F}$ is radial and $\mathbf{F} \cdot d\mathbf{l} = 0$ whenever $d\mathbf{l}$ is perpendicular to $\mathbf{F}$. That leaves

$$\int_A^B \mathbf{F} \cdot d\mathbf{l} = \int_{r_A}^{r_B} \left(-\frac{GMm}{r^2}\right) dr = \frac{GMm}{r_B} - \frac{GMm}{r_A} \,. \tag{7.28}$$

The work–energy theorem of Equation (7.21) then becomes

$$\frac{GMm}{r_B} - \frac{GMm}{r_A} = \frac{1}{2} m v_B^2 - \frac{1}{2} m v_A^2 \,,$$

or

$$\frac{1}{2} m v_A^2 - \frac{GMm}{r_A} = \frac{1}{2} m v_B^2 - \frac{GMm}{r_B} \equiv E \,. \tag{7.29}$$

Now since the total energy, E, is a constant of the motion, the points A and B are completely arbitrary and so for a *general* point on the orbit, characterised by a radius r we have

$$E = \frac{1}{2} m v^2 - \frac{GMm}{r} = K + U(r) \,, \tag{7.30}$$

where the potential energy function, $U(r) = -GmM/r^2$.

So far we have focused on the work–energy theorem and the conservation of energy. What about the conservation of angular momentum, Equation (7.14)? One deduction follows immediately from the fact that the angular momentum is a vector quantity. In order for the vector $\mathbf{L} = \mathbf{r} \times \mathbf{p}$ to maintain its direction, the motion of a central force must be *planar*. This is because if $\mathbf{L}$ is to maintain its direction then the vectors $\mathbf{r}$ and $\mathbf{p}$ must always be perpendicular to $\mathbf{L}$ (but not necessarily to each other). Note that this conclusion would not be so obvious if an origin were chosen at some point distinct from the source point of the force. In particular, a choice of origin out of the plane of the motion would result in an $\mathbf{L}$ vector which precesses. Referring now to Figs. 7.7 and 7.8 which introduce ϕ, the angle between the radius vector $\mathbf{r}$ and the velocity vector $\mathbf{v}$, the magnitude of $\mathbf{L}$ is given by $L = mrv \sin\phi$. The direction of $\mathbf{L}$ is into the plane of the paper as indicated in Figure 7.8. It is very useful to use the angle ϕ to project the velocity $\mathbf{v}$ into its radial and tangential components. The *tangential* velocity is $v \sin\phi$, whilst the *radial* velocity or $\dot{r}$ is $v \cos\phi$ so that the kinetic energy may be written

$$K = \frac{1}{2} m v^2 = \frac{1}{2} m \left[(v \cos\phi)^2 + (v \sin\phi)^2 \right] \,. \tag{7.31}$$

Notice that the tangential velocity component, $v \sin\phi$, can be written in terms of L and r, since $L = mrv \sin\phi$. Another way to write Equation (7.31) is therefore

$$K = \frac{1}{2} m \dot{r}^2 + \frac{L^2}{2mr^2} \,. \tag{7.32}$$

Combining this with Equation (7.30) we now have

$$E = \frac{1}{2}m\dot{r}^2 + \frac{L^2}{2mr^2} - \frac{GmM}{r} . \tag{7.33}$$

To summarise, we started by trying to work out how a particle moves in three dimensions under the influence of a central gravitational force. A consequence of the conservation of angular momentum is that the particle is confined to a two-dimensional plane. A further consequence of the conservation of angular momentum is that one of the planar coordinate variables can be eliminated in the expression for the total energy, which can thus be expressed entirely in terms of the radial coordinate, r. Although we have not exactly reduced the motion to one dimension – this would only be the case if we had reduced everything to a single equation of motion for r – we *have* managed to express an important conserved quantity, the total mechanical energy, purely in terms of r and $\dot{r}$.

Looking at the right-hand side of Equation (7.33) a little more closely, the first term, $m\dot{r}^2/2$, is the **radial kinetic energy**, or the energy due to motion in the radial direction only. The second term, $L^2/2mr^2$, derives from the contribution to the kinetic energy due to the tangential motion, *but is a function of r only, and therefore looks like a potential energy term.* We can even calculate the radial force, F_r, that this term defines:

$$F_r = -\frac{d}{dr}\left(\frac{L^2}{2mr^2}\right) = \frac{L^2}{mr^3} , \tag{7.34}$$

an inverse cubic repulsive term. What does this term correspond to physically? If the particle were describing a circle then r is a constant ($= R$, say), and the velocity is completely tangential, so $L = mvR$. The force of Equation (7.34) in this case is given by mv^2/R. You may recognise this as the so-called 'centrifugal force' (not a term I like) experienced by a particle moving in a circle. However, calling it the centrifugal force has not really explained anything. Where the force really arises from is appreciated by recognising that in deciding to restrict attention only to the radial motion, and ignoring the angular motion, we are effectively moving into a *rotating frame of reference.* We will consider rotating frames in detail in Chapter 9, where the origin of 'centrifugal force' will be examined again, but suffice it to note here that as a rotating frame is accelerating it will be necessary, in order to apply Newton's laws, to include 'inertial' or 'fictitious' forces as we discussed in Section 1.7.[1]

The third term in Equation (7.33) is the true gravitational potential energy which gives rise to the inverse square attractive force. Since we now have two functions of r, why not combine them into a single total potential function $U^*(r)$ where

$$U^*(r) = \frac{L^2}{2mr^2} - \frac{GMm}{r} ? \tag{7.35}$$

A plot of this *pseudo*-potential function, for a body with the same mass and angular momentum as the Earth, is shown in Figure 7.9. The meaning of the pseudo-potential is that it is the potential seen by the radial motion, and we can now invoke the machinery of potential functions considered in Chapter 2 to deduce all the main features of gravitational

[1] The astrocadet, emerging groggy from the training centrifuge, says, 'Well centrifugal force felt pretty real to me!' The point is his body was constrained to move in a circle whilst his body fluid wanted to move in a straight line. It is the former that gives rise to his feeling of force, not the latter. The fluid only accelerates in the sense that it does so relative to the body, which itself is accelerating relative to an inertial frame.

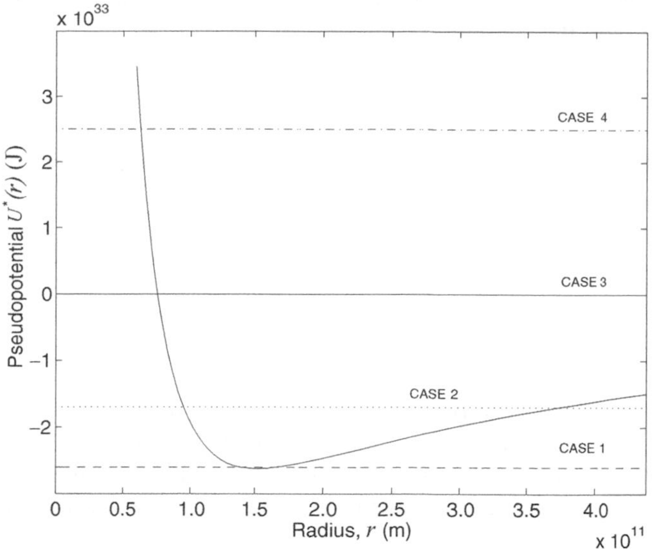

Figure 7.9 The psuedo-potential experienced by the Earth due to the Sun's gravitational influence.

orbits, without solving the equations of motion. This is another example of the qualitative approach to dynamical systems mentioned at the end of Section 2.8.

Once the masses and the angular momentum are identified then the pseudo-potential is fixed, and the only quantity left to specify the orbit is the total energy. Different total energies will yield different orbits. The total energy is a constant, independent not just of time but also of r, corresponding to a horizontal line on the psuedo-potential. Since the radial kinetic energy, $m\dot{r}^2/2$, is a positive quantity given by the difference between the total energy and the (pseudo-) potential energy, we know that a physical orbit must correspond to a total energy line that always lies above the $U^*(r)$ curve. With reference to the different total energy lines identified in Figure 7.9, the various cases are as follows.

Case 1: Minimum energy, $E = E_{\text{min}}$.

Here the total energy line touches the $U^*(r)$ curve tangentially at its minimum point. The only possible orbit is one that has a constant radius (= R), and the orbit is therefore a **circle**. Differentiating $U^*(r)$ yields the value of R:

$$\frac{d}{dr}U^*(r) = 0 \quad \Rightarrow r = \frac{L^2}{Gm^2M} \equiv R. \tag{7.36}$$

At this radius the energy is

$$U^*(r = R) = -\frac{G^2M^2m^3}{2L^2} = E_{\text{min}} \,. \tag{7.37}$$

Since $\dot{r} = 0$, the velocity is purely tangential, so $L = mvR$, and in this case R and E may be expressed in terms of the constant orbital speed v:

$$R = \frac{GM}{v^2}\,, \quad E_{\text{min}} = -\frac{GmM}{2R} = -\frac{1}{2}mv^2\,. \tag{7.38}$$

The orbital period, T, is just the circumferential distance travelled divided by v, so

$$T = \frac{2\pi R}{v} = 2\pi\frac{R^{3/2}}{\sqrt{GM}}\,, \tag{7.39}$$

that is $T^2 \propto R^3$.

Case 2: Bound orbit, $E_{\text{min}} < E < 0$.

Here the total energy is still negative and the physical orbit region, where $E \geq U^*(r)$, lies between two limiting radii, r_{min} and r_{max}. The planet will oscillate between these values throughout the orbit so that in general $\dot{r} \neq 0$. At the extreme radii, however, $E = U^*(r)$, $\dot{r} = 0$ and r_{min} and r_{max} may be calculated by solving the quadratic equation

$$E = U^*(r = r_{\text{max}\,/\,\text{min}}) = \left(\frac{L^2}{2mr^2_{\text{max}\,/\,\text{min}}}\right) - \frac{GMm}{r_{\text{max}\,/\,\text{min}}}\,. \tag{7.40}$$

The result is

$$r_{\text{max}\,/\,\text{min}} = -\frac{GmM}{2E}\left[1 \pm \sqrt{1 + \frac{2EL^2}{G^2m^3M^2}}\right]\,. \tag{7.41}$$

Remember that E is negative so that given the preceding minus sign, r_{min} and r_{max} are both positive. The most negative E can ever be is given by Equation (7.37) which shows that the square root in Equation (7.41) can never give an imaginary radius – phew! Detailed analysis achieved by solving the full equations of motion (for which you can consult more advanced textbooks) shows that the orbit in this case is an **ellipse** (see Figure 7.10). We have not proved that the orbit is elliptical, but the key features of the ellipse, its eccentricity,[2] ε, and semi-major axis, a, are easily derived. With coordinate origin at one focus of the ellipse,

$$r_{\text{max}} = a\,(1 + \varepsilon) \ \text{ and } \ r_{\text{min}} = a\,(1 - \varepsilon)\,, \tag{7.42}$$

so that by comparison with Equation (7.41)

$$a = -\frac{GmM}{2E} \ \text{ and } \varepsilon = \sqrt{1 + \frac{2EL^2}{G^2m^3M^2}}\,. \tag{7.43}$$

The Sun thus lies at the focus of the ellipse as shown in Figure 7.10. The radial extremities, r_{min} and r_{max}, are known respectively as *perihelion* and *aphelion*. The expression for the eccentricity can, with reference to conic section geometry, be used to classify all the orbits. The circular orbit of case 1, for example, corresponds to $\varepsilon = 0$. In the solar system Neptune's orbit is the most circular with $\varepsilon = 0.0040$, followed by Venus ($\varepsilon = 0.0068$) and then the Earth

[2]We are using the same symbol for the eccentricity as for the coefficient of restitution. However, planets rarely collide so there is little risk of confusion!

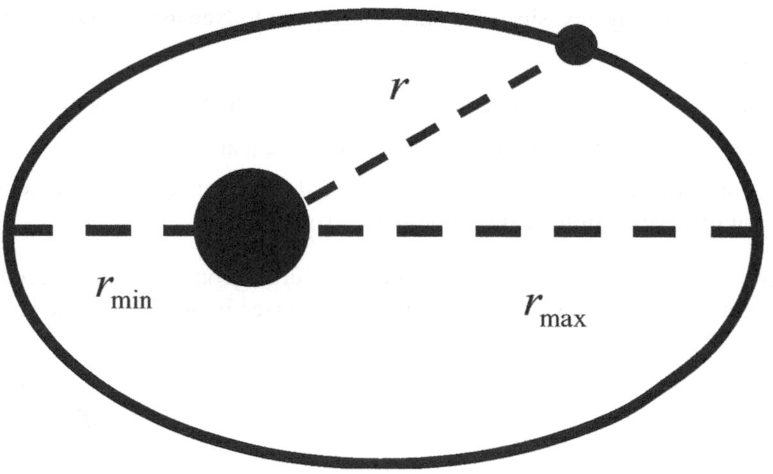

Figure 7.10 Planet orbiting the Sun. The Sun lies at one focus of an ellipse.

($\varepsilon = 0.0167$). Mercury's is the most eccentric with $\varepsilon = 0.2056$. Comets are characterised by highly eccentric orbits with ε approaching unity and beyond (see Case 4).

Case 3: Just unbound orbit, $E = 0$.

In this case the celestial body just manages to escape the gravitational influence of the Sun and escape to infinity. The closest approach is found by setting $E = 0$ and $\dot{r} = 0$ in Equation (7.33). The solution $r \to \infty$ corresponds to escape, whereas r_{min} is given by

$$r_{\text{min}} = \frac{L^2}{2GMm^2} \,. \tag{7.44}$$

Setting $E = 0$ in the second Equation (7.43) shows that the orbital eccentricity, $\varepsilon = 1$, which corresponds to a **parabola**.

Case 4: Unbound orbit, $E > 0$.

The body now escapes to infinity at which distance it has nonzero kinetic energy. The same procedure as for Case 2 of setting $\dot{r} = 0$ in Equation (7.33) and solving the resulting quadratic equation yields one unphysical negative radius, and the closest approach given by

$$r_{\text{min}} = \frac{GmM}{2E}\left[\sqrt{1 + \frac{2EL^2}{G^2m^3M^2}} - 1\right] = \frac{GmM}{2E}\left[\sqrt{1+\varepsilon} - 1\right] \,. \tag{7.45}$$

Here $\varepsilon > 1$ corresponds to a **hyperbola**. A comet with $\varepsilon > 1$ will visit the Sun just once before leaving the solar system forever.

7.6 Kepler's Laws

Kepler's three empirical laws concerning closed gravitational orbits have mainly historical significance, in that Newton's theory explained them comprehensively. They therefore have little status as laws of physics, since they are consequent to Newton's laws. We quote them here for completeness:

1. Planets move around the Sun in elliptical orbits with the Sun at a focus of the ellipse.
2. A planet's radius vector sweeps out equal areas in equal times.
3. If T is the orbital period then $T^2 \propto d^3$ where d is any scale distance associated with the orbit (e.g. perihelion, semi-major axis, perimeter, etc.).

The first law we have already covered, and needs no further comment.[3] The second law is a consequence of the conservation of angular momentum (see Figure 7.11). The area element,

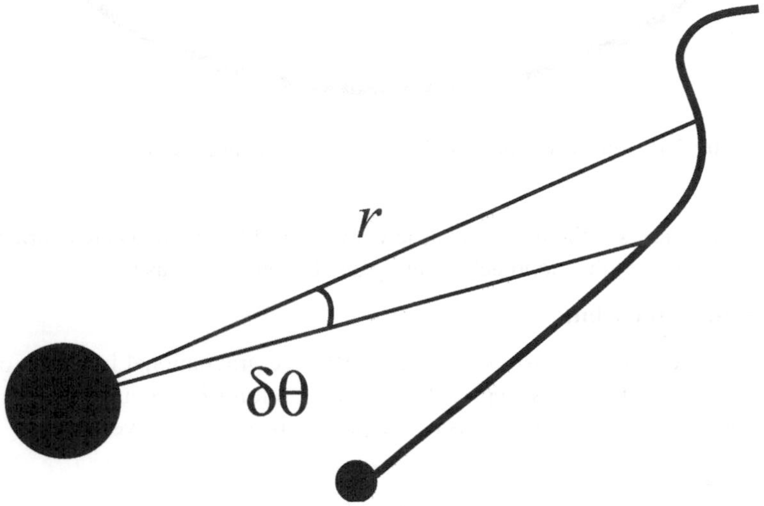

Figure 7.11 Infinitesimal area swept out by planetary orbit in time *dt* used in the calculation of Kepler's second law.

dA, swept out in time dt is given by $dA = r^2 d\theta/2$. The rate at which the area is swept out is therefore

$$\frac{dA}{dt} = \frac{1}{2} r^2 \frac{d\theta}{dt} .$$

However, the tangential velocity, $v_\theta = r d\theta/dt$, and the angular momentum, $L = mrv_\theta$, so

$$\frac{dA}{dt} = \frac{L}{2m},$$

which of course is constant.

The third law is quite interesting and shows how the orbits scale in space and time. It can be used to predict the period of a planet once its perihelion is known (or vice versa), given

[3]Except perhaps, the following curiosity. Readers old enough to remember the UK £1 banknote may remember its depiction of Newton against a background of a planet orbiting the Sun. The Sun was printed at the centre of the ellipse, not at a focus – Newton must have turned in his grave!

these quantities for the Earth. We have only proved the third law for the case of a circular orbit – Case 1 in Section 7.5 – but the more general form holds for any elliptical orbit. Any linear orbit measure can be chosen because all such distances are proportional to each other, with the same constant of proportionality for different ellipses.

7.7 Comments

We have focused exclusively on gravitational orbits as being the most important example of three-dimensional motion. However, the Coulomb interaction is also inverse square, and so what we have discussed also applies to electrostatic interaction. For unlike charges the interaction is attractive so that the formalism carries over, whereas for like charges it is repulsive and some reworking is necessary. It turns out that for a repulsive $1/r^2$ potential the orbit is always hyperbolic and no bound orbits are possible. Bound orbits occur for other force power laws. However, the inverse square law has a unique feature: only for this inverse power law are there stable *closed* orbits. A closed orbit is one that repeats itself after a certain period. A typical failure of a bound orbit to close is illustrated in Figure 7.12.

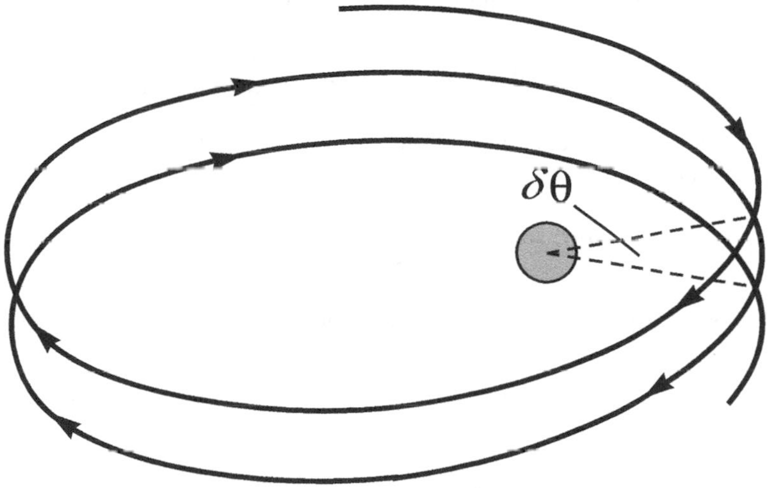

Figure 7.12 Failure of an orbit to close.

A notionally elliptical orbit is perturbed such that each time perihelion is reached, the radius vector is at a slightly different angular position, displaced by angle $\delta\theta$ say. Next time round the same will happen again so that with every period the perihelion position advances by $\delta\theta$. If a planet's perihelion is observed to advance it can mean two things: either the influence of other bodies in the solar system perturbs the ideal two-body case slightly or, more ominously, it can signal the failure of the gravitational inverse square law. The most famous example is the perihelion precession of Mercury. For this planet the perihelion radius advances by 5600 arc seconds per century. Of this, 5556 arc seconds are accounted for as the perturbational effects of the other planets. The tiny remainder of 46 arc seconds per century actually signalled the breakdown of Newton's law of gravitation, and provided one of the earliest quantitative tests of Einstein's general theory of relativity.

7.8 Summary

Section. 7.2. The work done by a force in moving along a path from point A to point B in three dimensions is given by

$$W = \int_{\text{path}} dW = \int_{A}^{B} \mathbf{F} \cdot d\mathbf{l} \,. \tag{7.4}$$

Section. 7.3. The **torque** $\mathbf{G} \equiv \mathbf{r} \times \mathbf{F}$ (Equation (7.7)), and the angular momentum $\mathbf{L} \equiv \mathbf{r} \times \mathbf{p}$, of a particle are related through

$$\mathbf{G} = \frac{d}{dt}(\mathbf{r} \times \mathbf{p}) \equiv \dot{\mathbf{L}} \,. \tag{7.8}$$

Section. 7.4. For **central forces** $\mathbf{F} = F(r)\hat{\mathbf{r}}$ and the work done is independent of the path taken. The work–energy theorem in three dimensions then applies:

$$W_{\text{AB}} = \int_{\text{A}}^{\text{B}} \mathbf{F} \cdot d\mathbf{l} = U(r_{\text{A}}) - U(r_{\text{B}}) = \frac{1}{2}mv_{\text{B}}^2 - \frac{1}{2}mv_{\text{A}}^2 \,. \tag{7.21}$$

where $U(r)$ is the potential function such that $F = -dU(r)/dr$ (Equation (7.18)). Energy conservation is then

$$\frac{1}{2}mv_{\text{A}}^2 + U(r_{\text{A}}) = \frac{1}{2}mv_{\text{B}}^2 + U(r_{\text{B}}) \equiv E \,. \tag{7.22}$$

Section. 7.5. For gravity $U(r) = -GMm/r$ and energy conservation can be written

$$E = \frac{1}{2}m\dot{r}^2 + U^*(r) \,, \tag{7.33}$$

where

$$U^*(r) = \frac{L^2}{2mr^2} - \frac{GMm}{r} \,, \tag{7.35}$$

defines the **pseudo-potential** (Figure 7.9). The constant total energy possessed by the pseudo-potential determines the orbit to be (1) circular ($E = E_{\text{min}}$), (2) elliptical ($E_{\text{min}} < E < 0$), (3) parabolic ($E = 0$), or (4) hyperbolic ($E > 0$).

Section. 7.6. Kepler's third law states: if T is the orbital period of a closed orbit then $T^2 \propto d^3$ where d is any scale distance associated with the orbit (e.g. perihelion, semi-major axis, perimeter, etc.).

7.9 Problems

7.1 A skater, mass 80 kg, on frictionless ice approaches point A (see Figure 7.13) with a velocity of $3\,\text{m s}^{-1}$. At A the skater grasps the free end of a taut massless rope which is attached to a pole at C by a frictionless ring. AC $= 2$ m. By pulling on the rope the skater follows a path to the point B where BC is 1 m. Calculate the initial angular momentum about C and the skater's final velocity. Account for the change in kinetic energy.

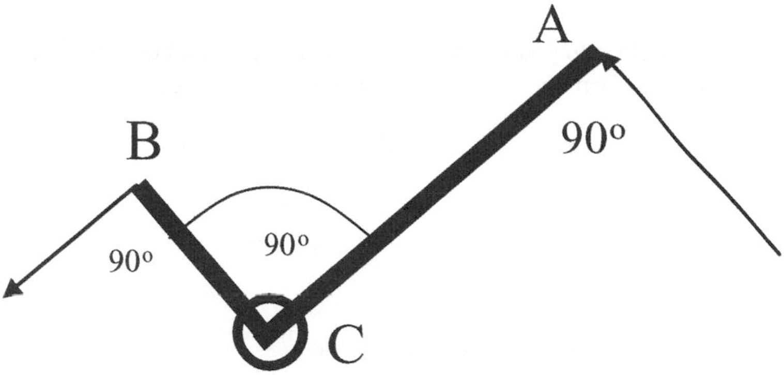

Figure 7.13 Figure for Problem 7.1.

Numerical data for Problems 7.2–7.5: $R_E = 6400\,\text{km}, g = 9.8\,\text{m s}^{-2}, G = 7 \times 10^{-11}\,\text{N m}^2\,\text{kg}^{-2}$, $M_{\text{Earth}} = 5.98 \times 10^{24}\,\text{kg}$, $M_{\text{Sun}} = 2 \times 10^{30}\,\text{kg}$.

7.2 A satellite moves in a circular orbit in the Earth's equatorial plane. Seen from the Earth, it appears to be stationary. Find the radius of its orbit. How many satellites are needed to survey every point on the equator?

7.3 The Moon's mass and radius are $0.012M_E$ and $0.27R_E$ (the subscript E refers to the Earth). For Jupiter the corresponding figures are $318M_E$ and $11R_E$. Find in each case the gravitational acceleration at the surface and the escape velocity.

7.4 Halley's comet is in an elliptical orbit about the Sun with an eccentricity of 0.97 and a period of 76 years. Calculate the distance in astronomical units (1 A.U. = Sun to Earth distance), of the comet from the Sun at perihelion ($= r_{\text{min}}$) and at aphelion ($= r_{\text{max}}$). (Hint: Use Kepler's third law.)

7.5 The velocity of the Sun in orbit around the galaxy is estimated to be about $200\,\text{km s}^{-1}$. Assume a circular orbit with a radius of 8000 parsecs (the distance at which the Earth–Sun distance subtends one second of arc, approximately 3×10^{16} m). Estimate, assuming a spherical mass distribution, the number of solar masses contained within a radius of 8000 parsecs about the centre of our galaxy.

7.6 A particle of mass 2 kg moves along the trajectory defined by $\mathbf{r} = 3\cos(2t)\mathbf{i} + \sin(2t)\mathbf{j} - \sqrt{3}\sin(2t)\mathbf{k}$ metres. Calculate:

(a) the particle's velocity and acceleration;

(b) the force acting on the particle;
(c) the angular momentum of the particle about the origin;
(d) the torque acting on the particle about the origin.
(e) From your answers to the above, what kind of force is acting on the particle?
(f) Draw a sketch showing the angular momentum vector in relation to the coordinate axes.

It should now be clear to you that we are not working in the best coordinate system for this problem. Transform the components of **r** (i.e. x, y, z) according to

$$x_{\text{new}} = x_{\text{old}} \text{ (i.e.leave } x \text{ unchanged)}\,,$$

$$y_{\text{new}} = \tfrac{1}{2} y_{\text{old}} - \tfrac{\sqrt{3}}{2} z_{\text{old}}, \text{ and}$$

$$z_{\text{new}} = \tfrac{\sqrt{3}}{2} y_{\text{old}} + \tfrac{1}{2} z_{\text{old}}\,.$$

(g) Sketch the trajectory of the particle in the (new) x–y plane.
(h) Now calculate:
   (i) the change in kinetic energy of the particle between $t = 0$ and $t = \pi/4$ s;
   (ii) the work done by the force, that is $\int_{r_1}^{r_2} F dr$ by noting the change in the radial coordinate between these times.

7.7 Disposal of nuclear waste and its containment over hundreds of years poses a serious problem. Let us consider the possibility of sending it off on a rocket and dumping it, *on the Sun!* After the launch the rocket will be travelling at the Earth's orbital speed

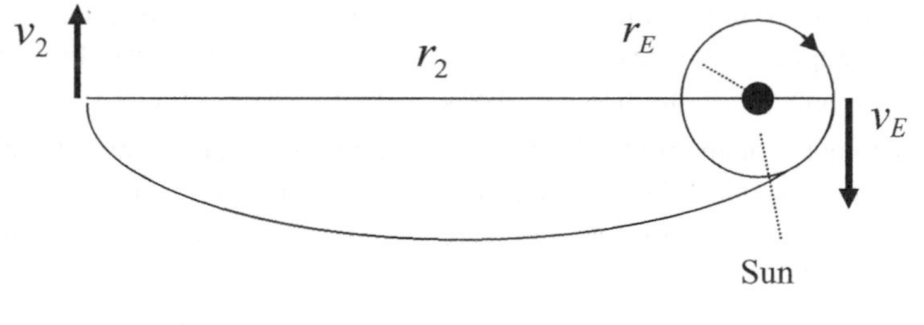

Figure 7.14 Figure for Problem 7.7

v_E. One way to ensure the rocket and its cargo reach the Sun is to fire the engines and bring the rocket to rest. It then has no angular momentum and can fall straight into the Sun. A much more economical way is to boost the rocket in the same direction to an increased speed v_1 so that it follows a new (elliptical) orbit with the Sun at a focus. At aphelion r_2 (furthest distance from the Sun), the rocket has a much lower speed v_2.

Here a reverse boost is applied stopping the craft in its orbit so that it then falls in to the Sun. Ignore any effect on the rocket due to the Earth's gravity (see Figure 7.14).

(a) Calculate the Earth's orbital speed, v_{E}, assuming a circular orbit about the Sun of radius $r_{\mathrm{E}} = 1.5 \times 10^{11}\,\mathrm{m}$ (1 year $\approx \pi \times 10^7\,\mathrm{s}$).

(b) Show that for the elliptical orbit, the speed at the furthest distance from the Sun, v_2, and at the closest distance, v_1, are related by: $v_2 = \beta v_1$ where $\beta = r_{\mathrm{E}}/r_2$.

(c) Using energy conservation, or otherwise, show that:

$$v_1^2 = \frac{2GM_{\mathrm{solar}}}{r_E(1+\beta)} \quad (M_{\mathrm{solar}} \text{ is the mass of the Sun})\,.$$

(d) Hence show that

$$v_1^2 = \frac{2v_E^2}{(1+\beta)}.$$

(e) Calculate the value of v_1 and of v_2 if $\beta = 0.1$.

(f) The speed boost required from the rocket engines in the direct operation is v_E. What is the total boost required in the second method (i.e. summing the boosts at both ends of the elliptical orbit)?

(g) Use the rocket equation (Equation (4.58)) and take a value $v_{\mathrm{ex}} = 3\,\mathrm{km\,s^{-1}}$ to show that a fuel saving of about 200-fold is obtained using the second method compared with the first.

8

Rigid Body Dynamics

8.1 Introduction

We noted in the previous chapter the complexity of Newton's second law applied in three dimensions. Matters are even more complicated when many particles are present, for then Equations (7.1) must be written down for each and every particle in the system. Along with Equations (7.1), the initial positions and velocities of every particle must also be specified. In principle, the evolution of the system is then known precisely. However, simply the number of equations and initial conditions excludes this 'brute force' approach as a practical method of calculation. In Chapter 4, we achieved some simplification through dividing two-body motion into external and internal motions, and to some extent this method can be applied to systems with more than two particles present. In this chapter we are going to simplify the billions of equations by imposing a *constraint* on the many-particle system. Suppose that all the particles comprising a system are locked together with rods that never buckle, stretch or otherwise deform. Consequently, the particles are no longer free to move independently of each other; if one moves, then *all* the others must respond instantaneously. Such an idealisation of a many-particle system is called a **rigid body**.

- A **rigid body** is defined as a system in which forces act so as to maintain the distances between its constituent particles.

A rigid body is indeed an idealisation. All extended bodies deform a little when subjected to forces, and of course the notion that any spatially separated particles can respond instantaneously to each other violates relativity. Nevertheless, providing we keep these limitations in mind, the rigid body concept is very useful. We must not get carried away, though. You might think that if all the bodies respond to each other at the same time, then effectively the dynamical problem reduces to that of a single particle. This is not so. A force applied to the centre of mass of a rigid body does indeed cause the body as a whole to accelerate. However, an external force acting through any other point exerts a *torque* about the centre of mass that causes the body to *spin* (see Figure 8.1). The spinning motion maintains the spatial separations, as required for a rigid body, but has no analogue for single-particle systems. Nevertheless, the problem *has* been effectively reduced to describing the centre-of-mass motion and the twisting effect induced by external forces and torques. With respect to the potential complexity of a many-particle system, the rigid body constraint is very severe. Our aim in this chapter is to analyse the residual dynamics consistent with this constraint.

Classical Mechanics – Second Edition From Newton to Einstein: A Modern Introduction Martin W. McCall

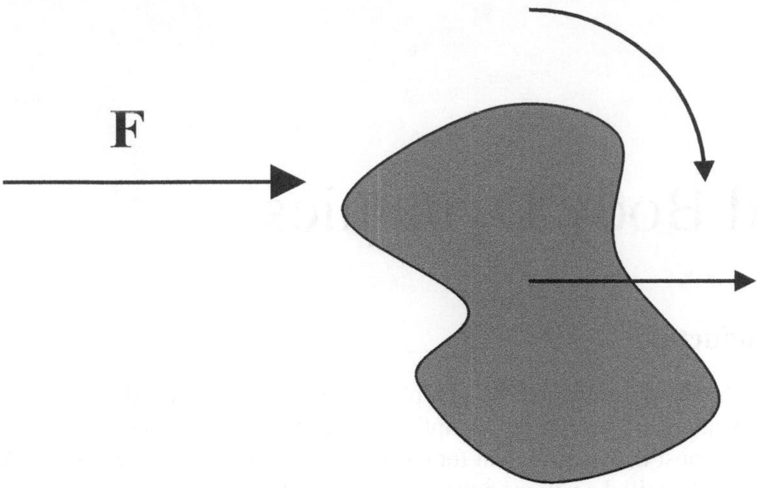

Figure 8.1 A free rigid body translating and rotating in response to an external force.

8.2 Torque and Angular Momentum for Systems of Particles

For a single particle we had (Equation (7.8)) $\mathbf{G} = d\mathbf{L}/dt$, and our first task is to generalise this to a system of particles. Figure 8.2 shows an arbitrary system, with as yet no rigid body constraint being applied. Forces acting on a particular particle are of two kinds. The *external* forces acting may be summed to give $\mathbf{F}_1^{\text{ext}}$, the total external force acting on particle 1, for example. Otherwise, there are the forces due to other particles in the system, where $\mathbf{F}_{12}$ denotes the force on particle 1 due to particle 2. The total torque on particle 1 is given by

$$\mathbf{G}_1 = \mathbf{r}_1 \times \mathbf{F}_1^{\text{ext}} + \mathbf{r}_1 \times \mathbf{F}_{12} + \mathbf{r}_1 \times \mathbf{F}_{13} + \cdots . \tag{8.1}$$

Similarly for mass 2:

$$\mathbf{G}_2 = \mathbf{r}_2 \times \mathbf{F}_2^{\text{ext}} + \mathbf{r}_2 \times \mathbf{F}_{21} + \mathbf{r}_2 \times \mathbf{F}_{23} + \cdots . \tag{8.2}$$

Now calculate the **total torque** as $\mathbf{G}_{\text{total}} = \mathbf{G}_1 + \mathbf{G}_2 + \cdots$. Evidently this sum will involve terms like

$$\mathbf{r}_1 \times \mathbf{F}_{12} + \mathbf{r}_2 \times \mathbf{F}_{21} ,$$

but by Newton's third law, $\mathbf{F}_{12} = -\mathbf{F}_{21} = \mathbf{f}$, say, so the contribution of such terms is

$$(\mathbf{r}_1 - \mathbf{r}_2) \times \mathbf{f} .$$

However, $(\mathbf{r}_1 - \mathbf{r}_2)$ is parallel to $\mathbf{f}$, so the cross product for these terms vanishes. The total torque acting on the system is therefore

$$\mathbf{G}_{\text{total}} = \mathbf{r}_1 \times \mathbf{F}_1^{\text{ext}} + \mathbf{r}_2 \times \mathbf{F}_2^{\text{ext}} + \cdots = \frac{d}{dt}(\mathbf{r}_1 \times \mathbf{p}_1) + \frac{d}{dt}(\mathbf{r}_2 \times \mathbf{p}_2) + \cdots , \tag{8.3}$$

where we have used Newton's second law to replace each external force by the rate of change of momentum it induces, and taken the time derivative outside the cross product, using the

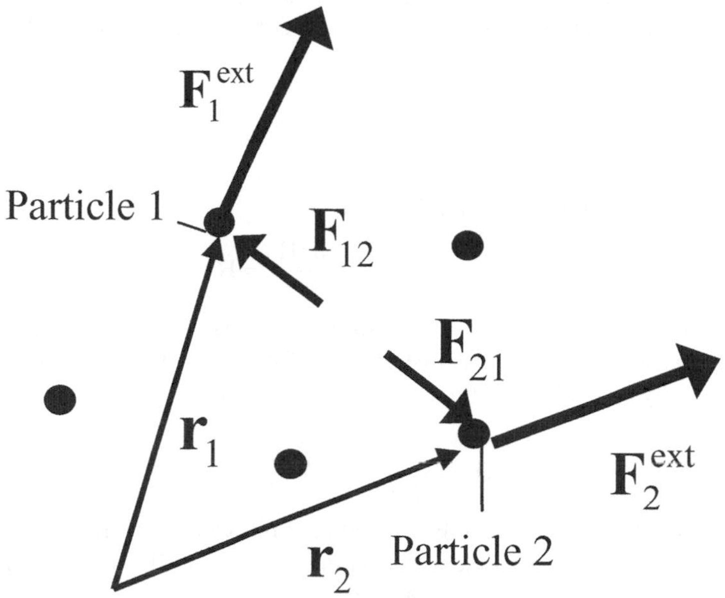

Figure 8.2 A system of particles.

same reasoning as given just after Equation (7.7). Defining the **total angular momentum** as

$$\mathbf{L}_{\rm total} = \mathbf{r}_1 \times \mathbf{p}_1 + \mathbf{r}_2 \times \mathbf{p}_2 = \mathbf{L}_1 + \mathbf{L}_2 + \cdots , \tag{8.4}$$

then Equation (8.3) can finally be written as

$$\mathbf{G}_{\rm total} = \frac{d\mathbf{L}_{\rm total}}{dt} . \tag{8.5}$$

The conclusion is that the *total* torque acting on the system is equal to the rate of change of the *total* angular momentum of the system. The result is *not* trivial. Would you have raised an eyebrow if I had just added the subscript 'total' to Equation (7.8)? It only works because all the internal torques cancel out. We note once again that, as for a single particle, $\mathbf{G}_{\rm total}$ and $\mathbf{L}_{\rm total}$ are in general dependent on the choice of origin.

8.3 Centre of Mass of Systems of Particles and Rigid Bodies

In Chapter 4 we defined the centre of mass for just two bodies (Equation (4.4)). Now we are dealing with many particles, the concept must be generalised. If there are N particles, each of mass m_i, located at position $\mathbf{r}_i$ ($i = 1, 2, \cdots , N$), then the centre-of-mass position, $\mathbf{R}$, is given by

$$\mathbf{R} = \frac{\sum_{i=1}^{N} m_i \mathbf{r}_i}{\sum_{i=1}^{N} m_i} = \frac{\sum_{i=1}^{N} m_i \mathbf{r}_i}{M} , \tag{8.6}$$

where M is the total mass of the system. Equation (8.6) is derived in a similar way to Equation (4.4), and represents a weighted average position over the system. If the total external force acting on the system is given by

$$\mathbf{F}_{\text{total}}^{\text{ext}} = \sum_{i=1}^{N} \mathbf{F}_i^{\text{ext}} , \tag{8.7}$$

then the centre of mass moves with acceleration

$$\mathbf{a}_{\text{CM}} = \frac{\mathbf{F}_{\text{total}}^{\text{ext}}}{M} = \frac{d^2\mathbf{R}}{dt^2} = \ddot{\mathbf{R}} . \tag{8.8}$$

The generalisation to a rigid body is fairly straightforward. The main complication is that a rigid body consists of a *continuous* distribution of matter, so that every particle mass m_i must be replaced by *mass element* dm, and the sum over the particles must be replaced by an integral:

$$\mathbf{R} = \frac{\int \mathbf{r}dm}{\int dm} = \frac{\int \mathbf{r}dm}{M} . \tag{8.9}$$

Over what limits is the integration to be carried out? Introducing the mass density ρ (which may depend on position) the mass element becomes $dm = \rho(\mathbf{r})dV$, where dV is the volume element occupied by dm, and Equation (8.9) becomes

$$\mathbf{R} = \frac{\int \mathbf{r}\rho(\mathbf{r})dV}{\int \rho(\mathbf{r})dV} = \frac{\int \mathbf{r}\rho(\mathbf{r})dV}{M} . \tag{8.10}$$

Evidently the integrals are to extend over the volume occupied by the body. This type of integration, where the domain is a *volume*, rather than an interval on the real number line, is called a **volume integral** and may be unfamiliar to you. It can, however, often be reduced to a familiar integration in a single variable, as will be the case in the examples we are going to consider below.

Incidentally, so far we have only presumed the system to consist of an extended distribution of matter without applying the rigid body constraint. Equation (8.8) therefore actually describes the centre of mass of *any* continuous distribution of matter, or, in other words, of any **fluid**.

Let us now look at a particular case of a rigid body in which the mass density is uniform ($\rho \neq \rho(\mathbf{r})$) and the shape of the body has a plane of symmetry as shown in Figure 8.3. Since mass is equally distributed above and below the plane, the centre of mass must lie within this plane. We prove this using Equation (8.10) and a little trick to avoid the complications of volume integration: suppose we know how the cross-sectional area, A, varies with height, z. Then $dV = A(z)dz$, and setting a coordinate origin somewhere within the plane, the z-component of the centre of mass is, from Equation (8.10), given by

$$z_{\text{CM}} = \frac{1}{M} \int_{-h/2}^{h/2} A(z)zdz , \tag{8.11}$$

where the body is presumed to lie between $z = -h/2$ and $z = h/2$. The integral in fact vanishes because $A(z)$ is an even function (i.e. $A(-z) = A(z)$), so $zA(z)$ is odd, and we are integrating over a symmetric interval. The centre of mass therefore lies in the plane of symmetry as claimed.

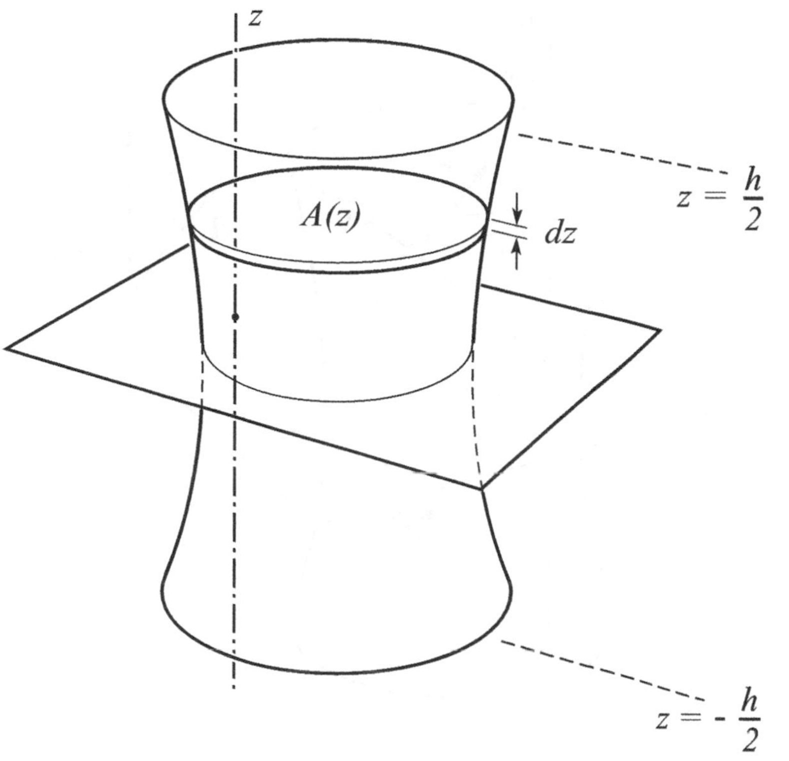

Figure 8.3 Calculation of the centre of mass of a uniform symmetric body.

8.4 Angular Momentum of Rigid Bodies

Calculation of the angular momentum for a rigid body is in general rather complicated and we will simplify the task by assuming the rigid body to be a uniform symmetric rigid body exactly as in Figure 8.3 with the axis of rotation being orthogonal to the plane of symmetry, but otherwise arbitrary (see Figure 8.4). Later on, in Section 8.7, we will relax this and give the general case.

The rigid body constraint implies that all points of the body rotate with the same angular speed ω. It turns out to be very useful to associate a direction with ω so that it becomes a little like a vector quantity. The only possible line defined is along the rotation axis, and we must make a choice between the two possible directions along this line. Although it does not matter which direction we choose as long as we are consistent, by convention the direction is chosen according to the *right-hand* rule, that is curl the fingers of the right hand in the same sense as the body is rotating (Figure 8.4), then the thumb is pointing along the direction of vector $\boldsymbol{\omega}$. The magnitude of $\boldsymbol{\omega}$ is of course the angular speed, ω. Now in general the axis of rotation will not pass through the centre of mass, though it will certainly pass through *some* point in the symmetry plane, which we will choose to be the coordinate origin O. A typical mass element of mass dm is located at position $\mathbf{r}$ and moves with velocity $\mathbf{v}$. Its contribution

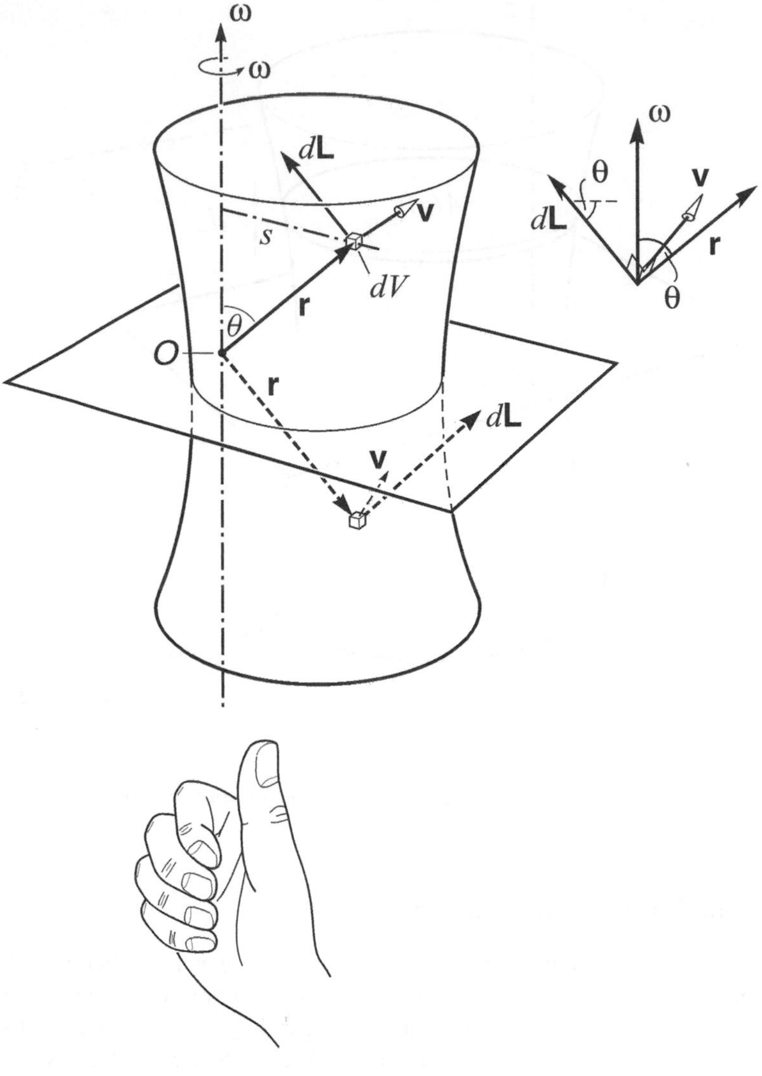

Figure 8.4 Rotating uniform symmetric rigid body.

to the body's angular momentum is therefore

$$d\mathbf{L} = \mathbf{r} \times (dm\mathbf{v}) = (\mathbf{r} \times \mathbf{v})\, \rho dV\ . \tag{8.12}$$

The three vectors in Equation (8.12) are shown together to the right of the main figure in Figure 8.4. Now for every such element there is another element with the same transverse coordinates, but situated the same distance *below* the centre of mass, as indicated. This complementary element (depicted with the broken vectors in the figure) has a similar angular momentum except that its horizontal component will be reversed. Therefore as a result of the choice of origin, the horizontal component of the *total* angular momentum is zero. The magnitude of the total angular momentum is then found by summing the vertical components so that

$$\mathbf{L}_{\text{total}} = \int |d\mathbf{L}| \sin\theta \,, \tag{8.13}$$

where θ is the angle between $\mathbf{r}$ and $\boldsymbol{\omega}$. The direction of $\mathbf{L}_{\text{total}}$ is vertically upwards, or parallel to $\boldsymbol{\omega}$. (This is a feature of the special case of a uniform symmetric body – it is *not* true in general, as we shall see later.) As the body rotates, the speed of the mass element is given by $v = |\mathbf{v}| = \omega s$, where s is the distance from the axis ($s = |\mathbf{r}| \sin\theta$) so that

$$|d\mathbf{L}| = |\mathbf{r} \times \mathbf{v}| \rho dV = rv\rho dV = r\omega s\rho dV = \omega r^2 \sin\theta \rho dV \,. \tag{8.14}$$

Note that $\mathbf{r}$ is orthogonal to $\mathbf{v}$. Finally, the magnitude of the total angular momentum can be written as

$$L_{\text{total}} = \int \omega s^2 \rho dV = \omega I \tag{8.15}$$

where $I \equiv \int \rho s^2 dV$ is called the **moment of inertia** of the body. This is again a volume integral, but we can simplify it to an integral in one variable using the same method as in Section 8.3. Let $\mathrm{A}(s)$ be the area of the body at distance s from the rotation axis (this is a *different* area to that considered in Section 8.3) – see Figure 8.5. We can now set $dV = \mathrm{A}(s)ds$, in which case

$$I \equiv \int_{s_1}^{s_2} \mathrm{A}(s)\rho s^2 ds \,. \tag{8.16}$$

The range of the integration is from s_1 to s_2, chosen so that the whole body is covered. Although the area $A(s)$ is shown as a planar area, if the rotation axis passes through the centre of mass, then it is invariably more convenient to have $A(s)$ represent the cylindrical area surrounding the axis at radius s. When the moment of inertia about such an axis is either known a priori or easily calculated ($= I_{\text{CM}}$, say), then there is a useful result for calculating the moment of inertia about any parallel axis. The distance s from a general axis to the mass element is then the sum of D_{CM}, the distance to the centre of mass, and D, the distance from the axis through the centre of mass to the mass element (see Figure 8.6). Therefore

$$s = D_{\text{CM}} + D, \quad ds = dD \,, \tag{8.17}$$

after which change of variables the moment of inertia about the general axis, I, is given by

$$\begin{aligned} I &\equiv \textstyle\int A(D)(D_{\text{CM}} + D)^2 \rho dD \,, \\ &= \textstyle\int A(D) D_{\text{CM}}^2 \rho dD + \int 2A(D) D_{\text{CM}} D \rho dD + \int A(D) D^2 \rho dD \,, \end{aligned} \tag{8.18}$$

where the limits have been omitted being understood to be such as to cover the body. Since D_{CM} is just the constant distance between the axes we can conclude that:

1. the first term on the right equals MD_{CM}^2;
2. the second term is proportional to the centre of mass expressed in terms of D. However, the centre of mass passes through $D = 0$, so this term vanishes;
3. the third term is the moment of inertia about the centre of mass, I_{CM}.

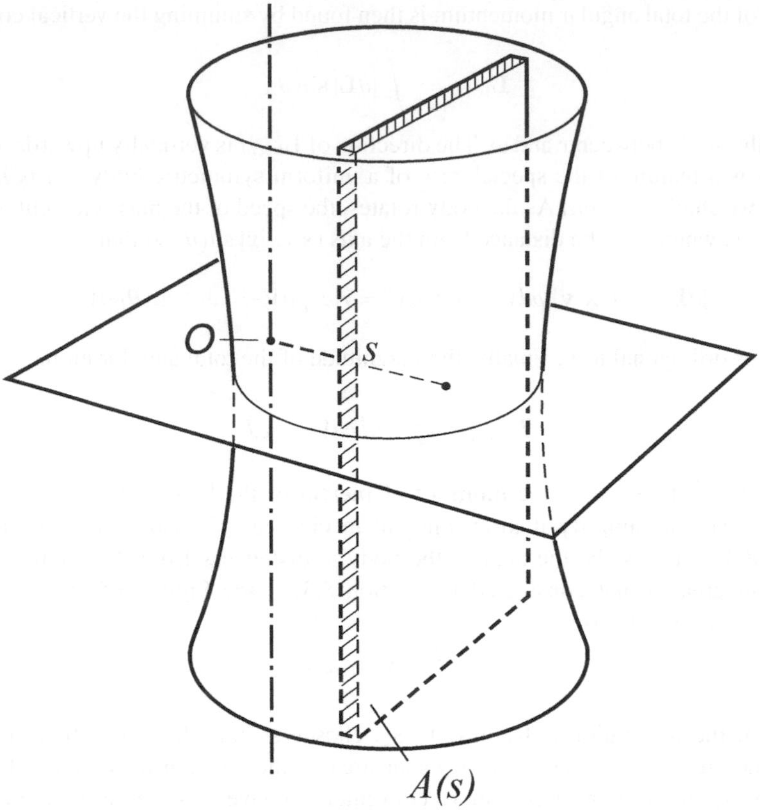

Figure 8.5 Illustrating $A(s)$, the area of the rigid body at distance s from the axis.

Putting these facts together we have finally

$$I = MD_{\mathrm{CM}}^2 + I_{\mathrm{CM}} \, , \tag{8.19}$$

a result known as the **Parallel Axes Theorem.**

Since linear momentum is mass times velocity, Equation (8.15) shows us that the moment of inertia plays a similar role in angular motion as does mass in linear motion. Unlike mass, however, I depends on the way the mass *is distributed*. We will examine some illustrative examples below.

Now that the total angular momentum has been calculated for a class of rotating symmetric bodies, the angular equation of motion (Equation (8.5)) becomes

$$\mathbf{G}_{\mathrm{total}} = \frac{d\mathbf{L}_{\mathrm{total}}}{dt} = I\frac{d\omega}{dt} \, , \tag{8.20}$$

which is the angular form of Newton's second law: Torque = Moment of Inertia × Angular Acceleration.

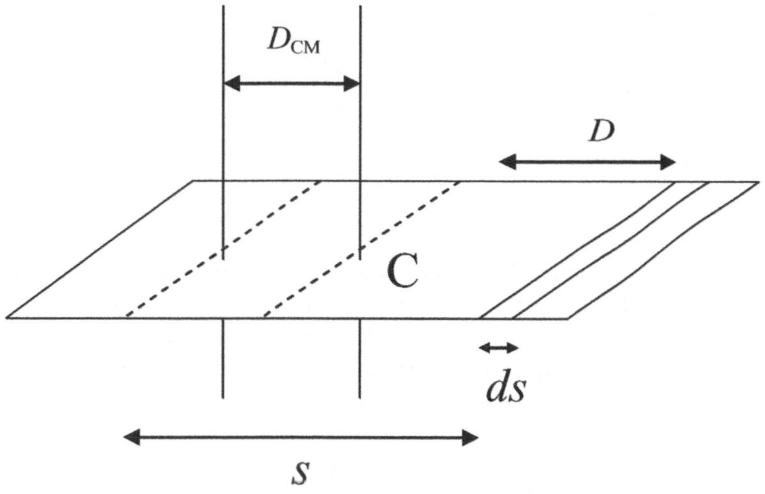

Figure 8.6 Geometry for the Parallel Axes Theorem; Point C is the centre of mass.

Example: Calculate the moment of inertia of a uniform bar of length l, cross-sectional area, A, rotating about an axis through its centre of mass, perpendicular to its length (see Figure 8.7).

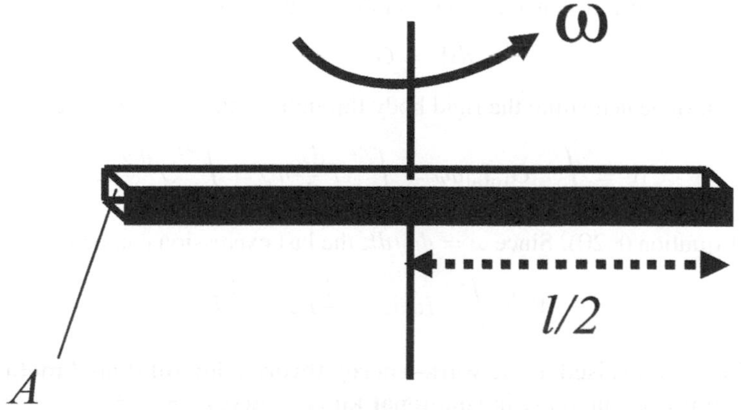

Figure 8.7 Uniform bar rotating about its centre of mass.

From Equation (8.16)

$$I = \int_{-l/2}^{l/2} s^2 \rho A dx = \rho\, A \frac{s^3}{3}\bigg|_{-l/2}^{l/2} = \rho A \frac{l^3}{12}\,. \tag{8.21}$$

However, since the rod is uniform $\rho = M/Al$ so

$$I = \frac{1}{12} M l^2\,. \tag{8.22}$$

Making the gross simplification that a spinning ice-skater may be modelled as consisting only of her outstretched arms (she is so thin that the contribution of her torso and legs is negligible!), Equation (8.21) may be used to see how she can increase her angular speed from ω_1 to ω_2 without any torque being applied. If her outstretched arms are separated by l_1 tip to tip when extended, and l_2 when pulled in, then conservation of angular momentum requires

$$\frac{1}{12} M l_1^2 \times \omega_1 = \frac{1}{12} M l_2^2 \times \omega_2\,, \tag{8.23}$$

or

$$\omega_2 = \left(\frac{l_1^2}{l_2^2}\right) \omega_1\,. \tag{8.24}$$

If the arms are pulled in reducing their length by a factor of two, then the angular speed increases by a factor of four.

8.5 Kinetic Energy of Rigid Bodies

In Section 7.3, when we introduced the notion of torque, it was related to the notion of a 'turning effect' induced by a force such that the work increment done in turning a particle through a small angle could be expressed as

$$dW = \text{Torque} \times \text{Angle Turned} = G \times d\phi\,. \tag{8.25}$$

Now that we have an expression for the total torque exerted on a system of particles, and a rigid body in particular, it is natural to generalise Equation (8.25) to

$$dW = G_{\text{total}} \times d\phi\,. \tag{8.26}$$

The total work done in turning the rigid body through a finite angle is therefore

$$W = \int_{\phi_1}^{\phi_2} G_{\text{total}} d\phi = \int_{\phi_1}^{\phi_2} I \frac{d\omega}{dt} d\phi = \int_{\omega_1}^{\omega_2} I \frac{d\phi}{dt} d\omega\,, \tag{8.27}$$

after using Equation (8.20). Since $\omega = d\phi/dt$, the last expression can be integrated to give

$$W = \int_{\omega_1}^{\omega_2} I\omega d\omega = \frac{1}{2} I\omega_1^2 - \frac{1}{2} I\omega_2^2\,. \tag{8.28}$$

This equation is recognised as the **work–energy theorem for rotational motion**, the right-hand side defining the increase in **rotational kinetic energy** ($= I\omega^2/2$). When a force acts on a rigid body and increases both its linear and angular momentum, Equation (8.28) must be combined with the linear work–energy theorem to calculate the total work done (see Problem 8.3).

8.6 Bats, Cats, Pendula and Gyroscopes

The best way to see how the formalism developed above works is to see it in operation in a number of specific examples. This section will consider two kinds of examples, those where no external torque acts (cats and bats), and those where gravity exerts a torque about the body's centre of mass (pendula and gyroscopes). In the former case angular momentum is conserved, whilst in the latter we will be using the angular equation of motion (Equation (8.20)).

1. Cricket Bats

Consider a ball of mass m striking a vertically held cricket bat of mass M (approximated as a uniform rod of length L) as shown in Figure 8.8. The ball impacts at the point x (as shown in the diagram) with velocity u. The bat is held lightly so that at the instant of the ball's impact the bat may be considered free. After the collision the ball moves with velocity v, without changing direction.

(a) Show that the moment of inertia of the bat about an axis perpendicular to the bat's length, passing through its centre of mass is $ML^2/12$. The bat's angular speed about the bat's centre of mass is ω and its centre-of-mass velocity after the collision is V.

(b) Write down the equations of conservation of angular and linear momentum of the system in terms of m, M, u, v, x, L and ω.

(c) By noting that the angular motion about the bat's centre of mass induces a linear velocity of $-\omega L/2$ at the top of the bat, show that the net velocity of the top of the bat is zero if $x = 2L/3$.

This problem involves a collision of a particle and a rigid body. Forces act during the short period of the collision, but are otherwise absent, so the conservation of linear and angular momentum *both* apply. Part (a) is just the rigid rod calculation at the end of Section 8.4 (Equation (8.22)). For part (b) we have for linear momentum conservation

$$mu = mv + MV \,. \tag{8.29}$$

Taking an origin at the bat's centre of mass, the conservation of angular momentum is

$$mu\left(x - \frac{L}{2}\right) = mv\left(x - \frac{L}{2}\right) + I_{\text{bat}}^{\text{CM}} \times \omega = mv\left(x - \frac{L}{2}\right) + M\frac{L^2}{12} \times \omega \,. \tag{8.30}$$

Now after the collision the bat is moving to the right, but is also rotating counterclockwise (say), so that as far as the top of the bat is concerned, these motions partially compensate each other. In fact the angular motion induces a velocity of $\omega L/2$ to the left, whilst the bat's centre-of-mass velocity to the right after the collision is V. The bat handle's velocity after the collision is therefore zero if $\omega L/2 = V$. Using this condition in Equation (8.30) and combining with Equation (8.29) we find

$$m(u - v)\left(x - \frac{L}{2}\right) = MV\left(x - \frac{L}{2}\right) = M\frac{L^2}{12} \times \frac{2V}{L} = M\frac{L}{6}V \,. \tag{8.31}$$

Solving Equation (8.31) for x yields $x = 2L/3$ as required.

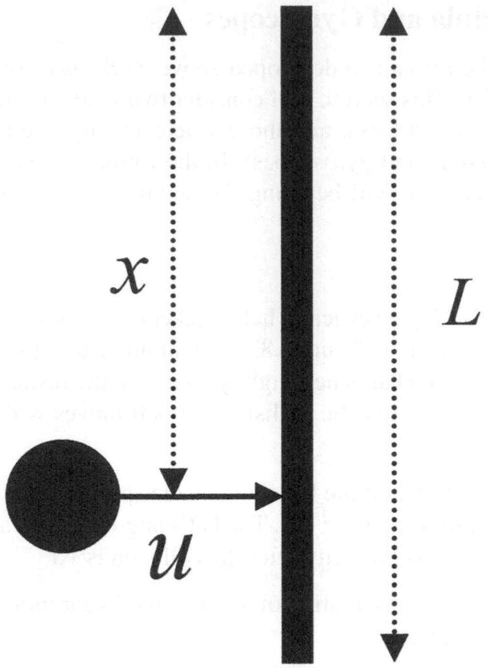

Figure 8.8 A ball striking an idealised cricket bat.

The point calculated is the most comfortable place for the batsman to hit the ball as at this point his hands will not feel any recoil of the bat. This optimum impact point is known as the **centre of percussion**.

2.Cats

Our next example is not strictly a single rigid body, but the conservation of angular momentum can still be used to solve this intriguing problem. Have you ever wondered how it is possible for a cat to always land on its feet even when it is dropped feet uppermost? Superficially, the cat's ability to turn through 180° without any external torque being applied would appear to violate the conservation of angular momentum.

One scheme through which the feat is possible is illustrated schematically in Figure 8.9. The cat is modelled as two counter-rotating cylinders connected by conical caps as shown which roll against each other without slipping. The two capped cylinders can separately be considered as rigid, so that their respective rotations can be represented by the angular velocity vectors $\boldsymbol{\omega}_1$ and $\boldsymbol{\omega}_2$ which have the same magnitude ($= \omega$) but different directions as shown. The total angular velocity, $\boldsymbol{\omega}_{\text{total}}$, is therefore directed horizontally and of magnitude $2\omega \cos\theta$, where θ is the angle at which the cylinders are inclined with respect to the horizontal. Since there is a symmetry plane vertically through the middle of the composite body (ignoring the 'catty' details), the resultant angular momentum vector is parallel to the resultant angular velocity. Since there is no external torque, the total angular momentum must always be zero. The *internal* angular momentum, generated by the twisting cylinders, must

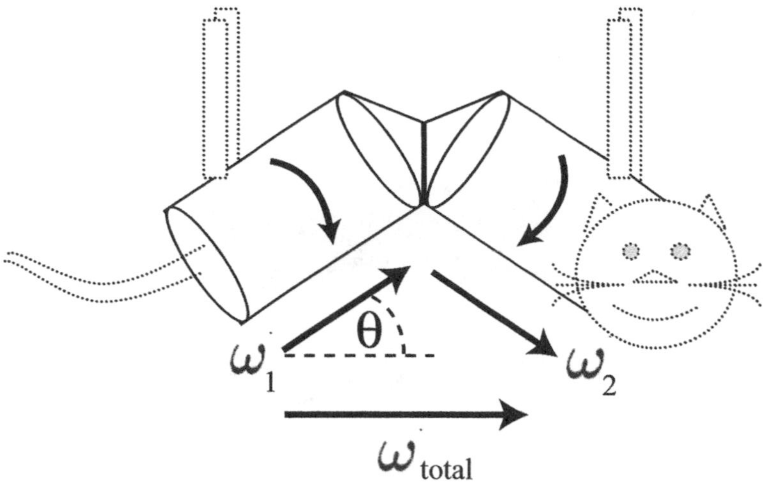

Figure 8.9 A schematic 'cat'.

be compensated for by an oppositely directed angular momentum vector causing the *body as a whole* to rotate in the opposite direction. The total angular momentum remains zero as there is no external torque. If the cat is dropped from height h (approximating the cat as a particle for this rough calculation) then it has time $t = \sqrt{2h/g}$ to turn through 180° before reaching the ground. Hence

$$2\omega t \cos\theta = 2\omega\sqrt{\frac{2h}{g}}\cos\theta = \pi\,. \tag{8.32}$$

The cat can complete the manoeuvre dropped from as little as 0.5 m (It is actually more stressful for the cat the lower it is dropped). For $\theta = 30°$, say, Equation (8.32) gives $\omega = 5.7$ rads^{-1}. Figure 8.10 illustrates a practical demonstration.[1]

3. A Compound Pendulum

A compound pendulum is one in which the mass is distributed along its length rather than in a bob. If it is fixed so that it can rotate about a point removed from the centre of mass, as in Figure 8.11, then gravity, which acts effectively through the centre of mass, will exert a torque about the axis. If the pendulum consists of a uniform rod of length l, and the axis is distant R from the centre of mass, then the moment of inertia about the axis is, by the parallel axes theorem and Equation (8.22)

$$I = I_{\text{CM}} + MR^2 = M\left(\frac{l^2}{12} + R^2\right)\,. \tag{8.33}$$

[1] Although the scheme undoubtedly correctly describes the general principles involved, the details may be slightly different. Over several trials (with two different cats!) the fore and hind legs are observed to consistently rotate at different rates, with the fore paws landing first.

Figure 8.10 'Charlie' doing the feat (under the supervision of a vet!). Courtesy Imperial College TV Studio.

We can now calculate the motion of the rotating rod under gravity. The magnitude of the torque gravity exerts on the rod is given by

$$G_{\text{total}} = MgR\sin\phi\,. \tag{8.34}$$

The first thing we can do is to calculate the pendulum's maximum angular speed when it is released from rest at an initial angle ϕ_0. The work done by the torque in restoring the pendulum to the vertical is, according to the first equality in Equation (8.27)

$$W = -\int_{\phi_0}^{0} MgR\sin\phi d\phi = MgR\left(1-\cos\phi_0\right)\,. \tag{8.35}$$

The minus sign in Equation (8.35) arises because the torque acts in the opposite angular direction to the direction of increasing ϕ. Using the work–energy theorem of Equation (8.28) the angular speed when the pendulum is vertical is determined via

$$MgR\left[1-\cos\phi_0\right] = \frac{1}{2}I\omega^2 = \frac{M}{2}\left(\frac{l^2}{12}+R^2\right)\omega^2\,.$$

The angular speed when the pendulum passes through the vertical is therefore

$$\omega = \left[\frac{2gR(1-\cos\phi_0)}{(l^2/12+R^2)}\right]^{1/2}\,. \tag{8.36}$$

The angular equation of motion is

$$G_{\text{total}} = -MgR\sin\phi = \frac{d}{dt}\left(I\dot{\phi}\right)\,. \tag{8.37}$$

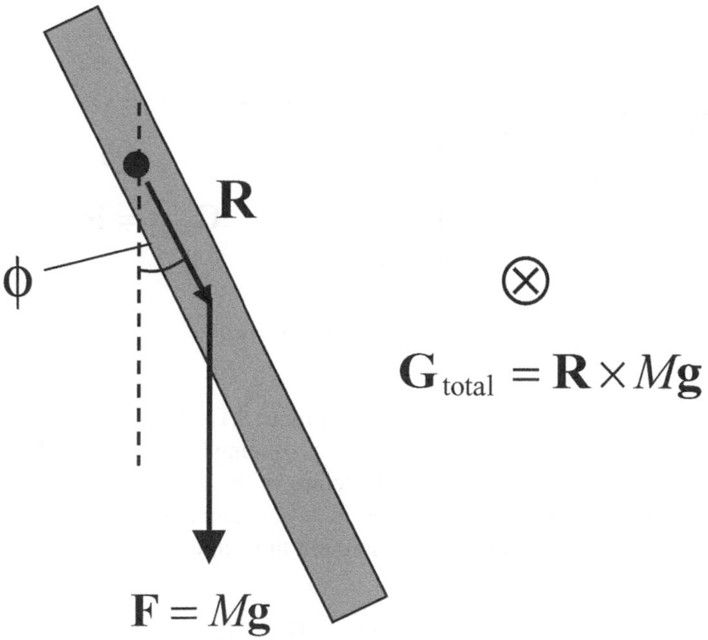

Figure 8.11 A compound pendulum.

As long as the pendulum's amplitude is not too large, $\sin\phi \approx \phi$ and we have

$$-MgR\phi \approx I\ddot{\phi} = M\left(\frac{l^2}{12} + R^2\right)\ddot{\phi}\,. \tag{8.38}$$

The motion is therefore simple harmonic, like a normal pendulum, except that the angular frequency is given by

$$(\text{Angular frequency}) = \left[g\bigg/\left(\frac{l^2}{12R} + R\right)\right]^{1/2}, \tag{8.39}$$

instead of $(g/l)^{1/2}$ for a standard bob pendulum of length l. If $R = l/2$ for example, then the angular frequency is $(3g/2l)^{1/2}$, or the period of the compound pendulum is $(2/3)^{1/2}$ that of the simple pendulum of the same length. If $R = 0$ then the period is infinite, since then the pendulum is balanced at the centre of mass and there is no torque.

4. A Gyroscope

In the examples considered so far we have presumed that the rotation of a body has been induced by the application of an external torque. However, it is possible for a torque to be applied to a body that is already spinning and possesses angular momentum. This case is interesting as the applied torque may not be in the same direction as the pre-existing angular momentum, so that its effect may be to produce a *new* rotation, rather than to increase the original angular velocity.

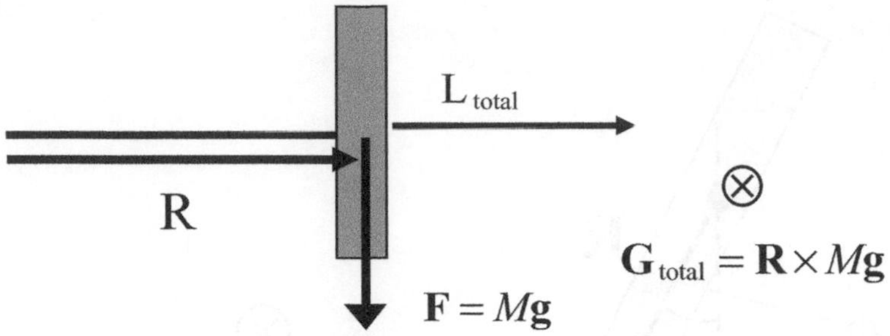

Figure 8.12 A gyroscope.

This is the basis of the gyroscope sketched in Figure 8.12. The flywheel is spinning with angular velocity as shown, the axle about which it is rotating being sufficiently light to regard the centre of mass as being located at **R**, the centre of the wheel. Gravity acts through the centre of mass to produce a torque $G_{\text{total}} = R \times Mg$, acting in the direction shown. In time dt the torque acts to produce a change in the angular momentum dL_{total} which, according to Equation (8.20), is given by

$$d\mathbf{L}_{\text{total}} = \mathbf{G}_{\text{total}}dt = (\mathbf{R} \times M\mathbf{g})dt \ . \tag{8.40}$$

The direction in which this change occurs is best seen by reverting to a plan view (see Figure 8.13). The angular increment, $d\psi$, through which the angular momentum vector has been

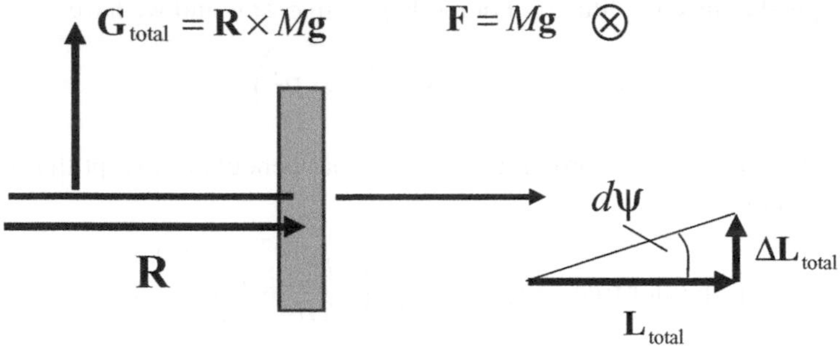

Figure 8.13 Plan view of a gyroscope.

steered is calculated as

$$d\psi = \frac{dL_{\text{total}}}{L_{\text{total}}} = \frac{G_{\text{total}}dt}{L_{\text{total}}} = \frac{MgR}{I\omega}dt \ , \tag{8.41}$$

where I is the moment of inertia of the flywheel about the axis. The effect of gravity is to cause the flywheel to **precess** with angular speed, Ω, given by

$$\Omega \equiv \frac{d\psi}{dt} = \frac{MgR}{I\omega} \ . \tag{8.42}$$

This precession about the vertical axis introduces a new upwards component to the angular momentum. There is no upward torque and therefore as the gyroscope precesses, the plane of rotation must also change to keep the vertical angular momentum zero.[2] These complications are avoided if we restrict attention to situations where the precession is much slower than the rate of flywheel rotation (i.e. $\Omega \ll \omega$). For a flywheel of radius a it turns out that $I = Ma^2/2$, so using Equation (8.42), the requirement for this analysis to be valid is that

$$\omega \gg \frac{\sqrt{2gR}}{a} . \tag{8.43}$$

8.7 General Rotation About a Fixed Axis

In Section 8.4 we had to go through some contortions to keep the maths simple. In this and the following sections we are going to generalise the earlier results so that we are no longer constrained to calculating the angular momentum of highly symmetric bodies. We will start by considering a set of masses connected by light rigid rods, rotating about a fixed axis, so that the angular velocity vector $\boldsymbol{\omega}$ points as shown in Figure 8.14. For simplicity we will choose an origin that lies somewhere on the axis of rotation.

A typical mass m_i at position $\mathbf{r}_i$ has velocity $\mathbf{v}_i$. As we saw in Section 8.4 the magnitude of $\mathbf{v}_i$ is $\omega r_i \sin\theta$, so that we have both the magnitude and the direction of $\mathbf{v}_i$ correct by writing $\mathbf{v}_i = \boldsymbol{\omega} \times \mathbf{r}_i$. Since the momentum of the mass element is $\mathbf{p}_i = m_i \boldsymbol{\omega} \times \mathbf{r}_i$, its angular momentum is given by

$$\mathbf{L}_i = m_i \mathbf{r}_i \times \mathbf{v}_i = m_i \left[\mathbf{r}_i \times (\boldsymbol{\omega} \times \mathbf{r}_i)\right] = m_i \left[r_i^2 \boldsymbol{\omega} - (\boldsymbol{\omega} \cdot \mathbf{r}_i)\, \mathbf{r}_i\right] , \tag{8.44}$$

where we have used the vector identity $\mathbf{a} \times (\mathbf{b} \times \mathbf{c}) = (\mathbf{a} \cdot \mathbf{c})\, \mathbf{b} - (\mathbf{a} \cdot \mathbf{b})\, \mathbf{c}$ (see Equation (A.4)). The total angular momentum is therefore given by the sum over all particles

$$\mathbf{L}_{\text{total}} = \sum_i m_i \left[r_i^2 \boldsymbol{\omega} - (\boldsymbol{\omega} \cdot \mathbf{r}_i)\, \mathbf{r}_i\right] . \tag{8.45}$$

Now by expanding $\left[r_i^2 \boldsymbol{\omega} - (\boldsymbol{\omega} \cdot \mathbf{r}_i)\, \mathbf{r}_i\right]$ into Cartesian components, the components of the total angular momentum of the body are given by

$$L_x = \sum_i m_i \left[\left(y_i^2 + z_i^2\right) \omega_x - x_i y_i \omega_y - x_i z_i \omega_z\right] , \tag{8.46}$$

$$L_y = \sum_i m_i \left[y_i x_i \omega_x - \left(z_i^2 + x_i^2\right) \omega_y - y_i z_i \omega_z\right] , \tag{8.47}$$

$$L_z = \sum_i m_i \left[z_i x_i \omega_x - z_i y_i \omega_y - \left(x_i^2 + y_i^2\right) \omega_z\right] . \tag{8.48}$$

We can conveniently express the above equations in matrix form as

$$\begin{bmatrix} L_x \\ L_y \\ L_z \end{bmatrix} = \begin{bmatrix} I_{xx} & I_{xy} & I_{xz} \\ I_{yx} & I_{yy} & I_{yz} \\ I_{zx} & I_{zy} & I_{zz} \end{bmatrix} \begin{bmatrix} \omega_x \\ \omega_y \\ \omega_z \end{bmatrix} \tag{8.49}$$

[2]The resultant motion is quite complicated involving an 'up-and-down' rocking known as nutation.

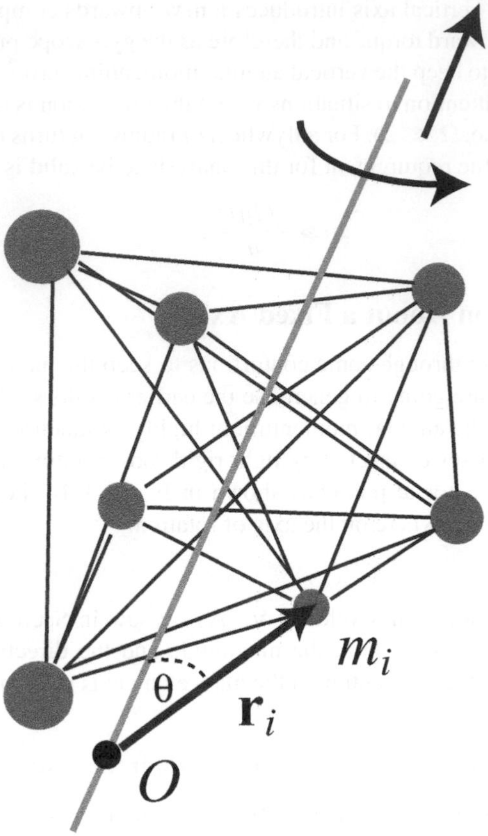

Figure 8.14 Set of masses connected by light rigid rods rotating about a fixed axis.

where

$$\begin{array}{lll} I_{xx} = \sum m_i \left(y_i^2 + z_i^2\right) , & I_{xy} = -m_i \sum x_i y_i , & I_{xz} = -m_i \sum x_i z_i , \\ I_{yx} = -m_i \sum m_i y_i x_i , & I_{yy} = m_i \sum \left(z_i^2 + x_i^2\right) , & I_{yz} = -m_i \sum y_i z_i , \\ I_{zx} = -m_i \sum z_i x_i , & I_{zy} = -m_i \sum z_i y_i , & I_{zz} = m_i \sum \left(x_i^2 + y_i^2\right) . \end{array} \quad (8.50)$$

Most compactly Equation (8.49) can be written as

$$\mathbf{L}_{\text{total}} = \mathbf{I}\boldsymbol{\omega} , \quad (8.51)$$

where $\mathbf{I}$ is called the **inertia tensor**, and the product $\mathbf{I}\boldsymbol{\omega}$ is interpreted as the product of a matrix on a vector – see the Appendix (Section A.4). If the system consists of a mass density distribution $\rho(\mathbf{r})$ occupying a volume V then summations may be replaced by integrals so that for example

$$I_{yz} = -\int_V \rho(\mathbf{r}) yz dV .$$

You can see from the definitions that for both discrete and continuous mass distributions $I_{xy} = I_{yx}$ etc., so that the inertia tensor is *symmetric*, a crucial property that will allow a great simplification later on. Just as in Section 8.4 the relationship between $\mathbf{L}_{\text{total}}$ and $\boldsymbol{\omega}$ is linear (cf. Equation (8.13)), although it is no longer the case that $\mathbf{L}_{\text{total}}$ is necessarily parallel to $\boldsymbol{\omega}$. Let us look at a simple illustrative example that explains why this is the case.

Take two equal point masses (m) connected by a rigid light rod (length l) and suppose that the system is rotating with constant angular velocity ω about the z-axis passing through the centre of mass and to which the rod makes an angle θ (see Figure 8.15).

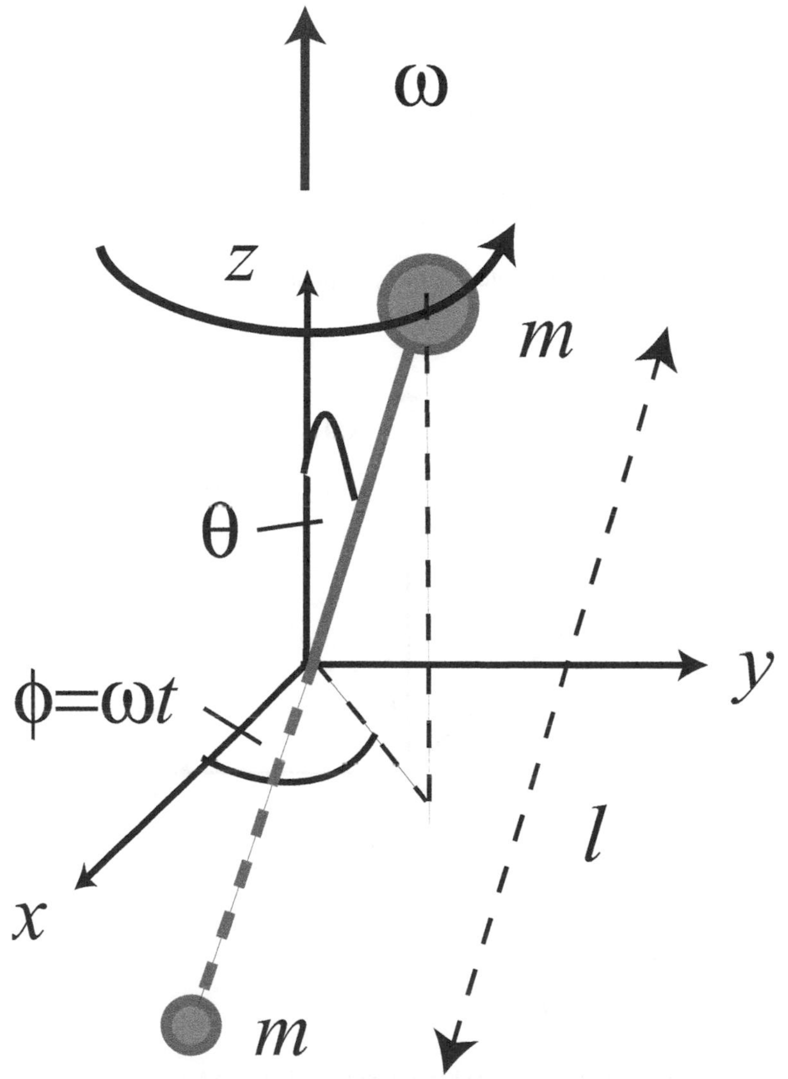

Figure 8.15 Two masses connected by a light rigid rod rotating about a fixed axis. The masses are considered to be particles.

If initially the masses lie in the x–z plane, then the position vectors of the two masses are $\mathbf{r} = \pm\frac{l}{2}(\sin\theta\cos\omega t, \sin\theta\sin\omega t, \cos\theta)$, and from Equation (8.50) we find

$$I_{xx} = \frac{ml^2}{2}\left(\sin^2\theta\sin^2\omega t + \cos^2\theta\right) , \tag{8.52}$$

$$I_{xy} = -\frac{ml^2}{2}\sin^2\theta\sin\omega t\cos\omega t , \tag{8.53}$$

$$I_{xz} = -\frac{ml^2}{2}\sin\theta\cos\theta\cos\omega t , \tag{8.54}$$

$$I_{yy} = \frac{ml^2}{2}\left(\sin^2\theta\cos^2\omega t + \cos^2\theta\right) , \tag{8.55}$$

$$I_{yz} = -\frac{ml^2}{2}\sin\theta\cos\theta\sin\omega t , \tag{8.56}$$

$$I_{zz} = \frac{ml^2}{2}\sin^2\theta . \tag{8.57}$$

When $t = \pi/2\omega$ the masses are located in the y–z plane at $\pm\frac{l}{2}(0, \sin\theta, \cos\theta)$ as shown in Figure 8.16.

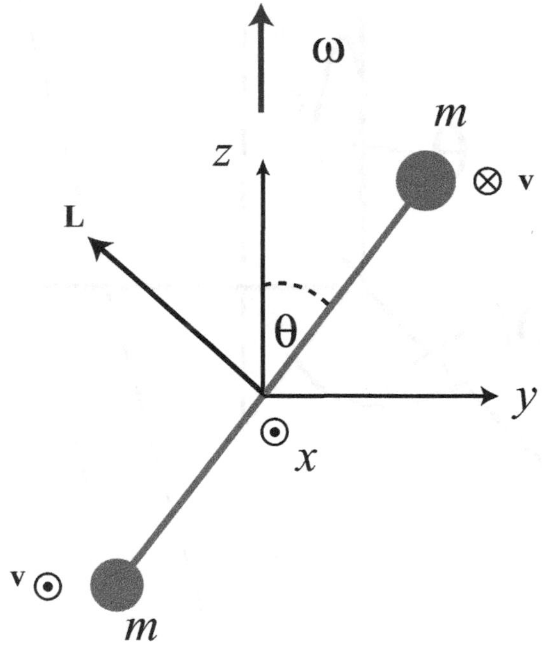

Figure 8.16 Rotating bar of Figure 8.15 at time $t = \pi/2\omega$.

Now imagine sitting to the right in the diagram squinting back along the y-axis towards the origin. You would see the upper mass moving to the right and the lower mass moving to the left. At this moment, therefore, *the masses are seen to rotate about the y-axis*. You can also see that if you were squinting up the z-axis from a position below the system, at this instant you would also see the masses rotating about the z-axis. The masses are actually describing circles about the z-axis, as indicated by the fact that $\boldsymbol{\omega}$ lies along the z-axis. That the masses are *instantaneously* seen to rotate about both the y-axis and the z-axis shows that at this instant the angular momentum vector has components in both the y and z directions. In fact $L_y = I_{yz}\omega = -\frac{m\omega l^2}{2}\sin\theta\cos\theta$ and $L_z = I_{zz}\omega = \frac{m\omega l^2}{2}\sin^2\theta$. So $\mathbf{L}$ is actually perpendicular to the rod, making an angle $(90^o - \theta)$ to the z-axis. As time progresses the angular momentum precesses about $\boldsymbol{\omega}$ (the z-axis). Note that the components of the inertia tensor are in general *not* constant, even when, as in this example, the angular velocity is. From Equation (8.51) this implies that the total angular momentum also changes. Since we also know that the rate of change of angular momentum equals the torque acting on the system (see Equation (8.5)), we deduce that there must be a torque present. Let us pursue this a little. Using Equations (8.49), (8.54), (8.56) and (8.57), and the fact that $\boldsymbol{\omega} = \omega\mathbf{k}$ we have

$$L_x = I_{xz}\omega = -\frac{ml^2}{2}\sin\theta\cos\theta\cos\omega t\ , \tag{8.58}$$

$$L_y = I_{yz}\omega = -\frac{ml^2}{2}\sin\theta\cos\theta\sin\omega t\ , \tag{8.59}$$

$$L_z = I_{zz}\omega = \frac{ml^2}{2}\sin^2\theta\ . \tag{8.60}$$

Therefore the time derivative of the angular momentum when $t = \pi/2\omega$ is given by

$$\frac{d\mathbf{L}}{dt} = \frac{ml^2\omega}{2}\sin\theta\cos\theta\mathbf{i} = \mathbf{G}\ . \tag{8.61}$$

How is this torque to be interpreted? Let us think about the forces acting on the masses as they rotate. Their positions are given by $\mathbf{r} = \pm l/2(\sin\theta\cos\omega t, \sin\theta\sin\omega t, \cos\theta)$, so that the force acting on the upper mass is $m\ddot{\mathbf{r}} = -m\omega^2 l/2(\sin\theta\cos\omega t, \sin\theta\sin\omega t, 0)$, and on the lower mass is $m\omega^2 l/2(\sin\theta\cos\omega t, \sin\theta\sin\omega t, 0)$. At $t = \pi/2\omega$ these forces are respectively $-m\omega^2 l\sin\theta/2\mathbf{j}$ and $m\omega^2 l\sin\theta/2\mathbf{j}$. These forces combine to exert a torque of magnitude $2\times(l/2)\cos\theta\times m\omega^2(l/2)\sin\theta = m\omega^2(l^2/2)\sin\theta\cos\theta$. When two equal and opposite forces act to produce a torque in this way, it is called a **couple**[3]. At $t = \pi/2\omega$ this couple acts in the $+x$ direction and is precisely the torque of Equation (8.61). So the torque of Equation (8.61) is to be interpreted as the couple that must be supplied at the axis in order to act against the system's propensity to 'flatten out' and lie in the x–y plane. Notice one final important point. If $\theta = 0$ or $\pi/2$ this torque is *zero*, and no there is no couple at the axis. In these orientations the system is 'balanced', with minimal wear on the bearings at the axis. We will now explore these special directions in more detail.

[3] See if you can prove that, unlike a general torque, a couple is independent of the choice of origin.

8.8 Principal Axes

The previous example showed us that when the system is rotating about certain axes it is balanced. From an engineering point of view it would certainly be of interest to know which directions have this property, as this will minimise wear on the bearings at the axis supporting the rotating system. For the previous example it is easy to see that if our wobbly bar is oriented in the x–y plane (i.e. $\theta = \pi/2$), then no turning forces need to be supplied at the axis to maintain the system at $\theta = \pi/2$. You can imagine balancing the bar on your finger and setting it spinning. In this example it is sort of 'obvious' that the rotation will feel balanced. However, what if we have some irregularly shaped body? In this section we will develop a powerful method to show how to find the special axes (they always exist) that support balanced, couple-free rotation. These special axes are called the **principal axes**.

Let us start with an arbitrary body at rest. Even though it isn't rotating yet, we can still calculate its inertia tensor according to Equation (8.50) and we end up with the numbers

$$\begin{bmatrix} I_{xx} & I_{xy} & I_{xz} \\ I_{yx} & I_{yy} & I_{yz} \\ I_{zx} & I_{zy} & I_{zz} \end{bmatrix} . \tag{8.62}$$

Since the body is stationary then all elements of the array are independent of time. Now imagine it happens that there exists some clever rotation of the coordinates such that in the new coordinates (labelled x_1, x_2, x_3) the array takes the simple diagonal form

$$\mathbf{I} = \begin{bmatrix} I_1 & 0 & 0 \\ 0 & I_2 & 0 \\ 0 & 0 & I_3 \end{bmatrix} , \tag{8.63}$$

where, for example, $I_3 = \sum_i m_i(x_{1i}^2 + x_{2i}^2)$. The elements I_1, I_2 and I_3 are still time-independent. Now set the system to rotate about the axis defined by the unit vector $\mathbf{e}_3$, so that in the new coordinate system we have $\boldsymbol{\omega} = \omega\mathbf{e}_3$. Even though the system is now rotating, the quantity $(x_{1i}^2 + x_{2i}^2)$ will remain constant for each mass element, because this is just the square of the element's fixed distance from the x_3-axis. The angular momentum will then be given by

$$\mathbf{L} = \mathbf{I}\boldsymbol{\omega} = I_3\omega\mathbf{e}_3 . \tag{8.64}$$

The angular momentum points in the same direction as the angular velocity, *and will be independent of time*, as both I_3 and the direction $\mathbf{e}_3$ are constant. Clearly the x_3-axis is a principal axis, and the moment of inertia about this axis, I_3, is called a **principal moment of inertia**. You can see that similar arguments would apply were we instead to set the system rotating about either the x_1- or x_2- axes. So if it is possible to find a transformation that takes the static-body inertia tensor to the form Equation (8.63), then we would have identified three principal axes, and three principal moments of the system. Now the remarkable thing is that it turns out that such a transformation is *always* possible. It doesn't matter how irregular the original object is – a potato, a chair, a car – for a given origin, a set of three orthogonal axes can always be found in which the static inertia tensor takes the form of Equation (8.63). The key point is that the inertia tensor defined by Equation (8.50) is *symmetric*, that is its elements satisfy $I_{xy} = I_{yx}$ etc. Now in Section A.7 of the Appendix it is shown that when a symmetric matrix operates on a cartesian vector, one can always find three orthogonal vectors for which the action is simply to multiply each vector by a constant. Once the directions associated with these vectors have been found, we can rotate the original coordinate axes to be aligned with these directions. In the new coordinate system the static inertia tensor takes the form of Equation (8.63). Alternatively, we could stick with the original coordinate system and express the principal directions in this system. The technique for finding the principal directions, given the elements of the inertia tensor, is described in Section A.6. In particular, to find the principal moments of inertia we must solve the equation

$$\det|\mathbf{I} - \lambda\mathbf{Id}| = 0\,, \tag{8.65}$$

for the possible values of λ, that are the *eigenvalues* of $\mathbf{I}$. Here $\mathbf{Id}$ is the identity operator (i.e. $\mathbf{Id}\,(\mathbf{v}) = \mathbf{v}$ for any vector $\mathbf{v}$). Since Equation (8.65) is a cubic equation in λ, there are, in general, three eigenvalues $\lambda = I_1, I_2, I_3$, that is three principal moments of inertia.

As a very simple example let us consider our two-mass wobbly bar when it lies in the y–z plane so that the two masses are located at $\pm\frac{l}{2}\,(0, \sin\theta, \cos\theta)$ and we have

$$\mathbf{I} = \frac{ml^2}{2}\begin{bmatrix} 1 & 0 & 0 \\ 0 & \cos^2\theta & -\sin\theta\cos\theta \\ 0 & -\sin\theta\cos\theta & \sin^2\theta \end{bmatrix}.$$

You can easily verify (or calculate using the method of Section A.6) that the eigenvalues and a set of normalised eigenvectors are given by

$$I_1 = \tfrac{1}{2}ml^2\,,\quad \begin{bmatrix} 1 \\ 0 \\ 0 \end{bmatrix}\;;\quad I_2 = \tfrac{1}{2}ml^2\,,\quad \begin{bmatrix} 0 \\ -\cos\theta \\ \sin\theta \end{bmatrix}\;;\quad I_3 = 0\,,\quad \begin{bmatrix} 0 \\ \sin\theta \\ \cos\theta \end{bmatrix}.$$

Thus when the bar lies in the y–z plane the x-axis is principal, with principal moment given by the usual $\frac{1}{2}ml^2$. Similarly the axis orthogonal to the rod lying in the y–z plane is also principal with moment $\frac{1}{2}ml^2$. Finally, the zero–eigenvalue solution is associated with the axis of the rod itself about which the masses, considered as particles, have no moment of inertia. The degeneracy[4] in the moments $I_1 = I_2$ implies that the choice of principal axes is not unique. Any orthogonal pair in the same plane as $[1, 0, 0]$ and $[0, -\cos\theta, \sin\theta]$ will suffice. This reflects the rotational symmetry.

If, as in this example, just two of the principal moments are identical, the system is called a **symmetric top**, whilst if it happens that all principal moments coincide, we have a **spherical top**. Thus the wobbly bar of Section 8.7 is a symmetric top, whilst the cube rotating about any axis through its centre is a spherical top, as we shall prove shortly.

8.9 Examples of Principal Axes and Principal Moments of Inertia

For a slightly more complicated example, consider the rod-connected three-mass system shown in Figure 8.17. Given the values of the masses and their indicated coordinates, we

[4]Degeneracy is said to occur when at least two eigenvalues are equal

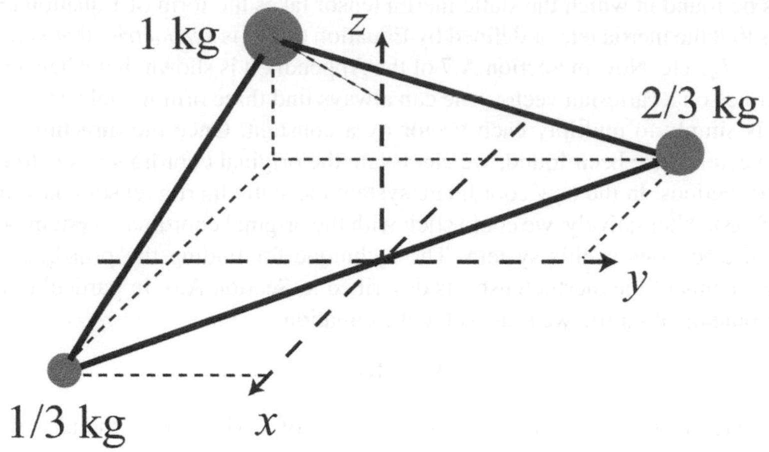

Figure 8.17 Three-mass system consisting of a mass 1/3 kg located at $(\mathbf{i} - \mathbf{j})$m, a mass 2/3 kg located at $(-\mathbf{i} + \mathbf{j})$ m, and a 1 kg mass located at $(-\mathbf{i} - \mathbf{j} + \mathbf{k})$ m.

find from Equation (8.50) that

$$\mathbf{I} = \begin{pmatrix} 3 & 0 & 1 \\ 0 & 3 & 1 \\ 1 & 1 & 4 \end{pmatrix} \text{ kg m}^2 . \tag{8.66}$$

The principal moments (eigenvalues) and principal axes (eigenvectors) are calculated straightforwardly (Section A.6) as

$$I_1 = 3 \text{ kg m}^2 \ , \ \mathbf{e}_1 = \mathbf{i} - \mathbf{j} \, , \tag{8.67}$$

$$I_2 = 2 \text{ kg m}^2 \ , \ \mathbf{e}_2 = \mathbf{i} + \mathbf{j} - \mathbf{k} \, , \tag{8.68}$$

$$I_3 = 5 \text{ kg m}^2 \ , \ \mathbf{e}_3 = \mathbf{i} + \mathbf{j} + 2\mathbf{k} \, , \tag{8.69}$$

where no attempt has been made to normalize the eigenvectors. If, from the origin, we were to pick any of the principal directions and set the system spinning, there would be no couple exerted on the axis and the system would feel balanced. We should emphasise that the choice of the point in the system at which to calculate the principal axes is important. If, in Figure 8.17, we were to displace all the masses by 1 m in the x direction and then 1 m in the y direction, their new locations would be 2**i** m, 2**j** m, and 1**k** m, and the inertia tensor would become

$$\mathbf{I} = \frac{1}{3} \begin{pmatrix} 11 & 0 & 0 \\ 0 & 7 & 0 \\ 0 & 0 & 12 \end{pmatrix} \text{ kg m}^2 . \tag{8.70}$$

The inertia tensor is now already diagonal (so we just read off the principal moments) and the principal axes coincide with the cartesian coordinate axes. Changing the point at which to calculate the principal axes changes both the principal moments of inertia *and* the orientation of the principal axes.

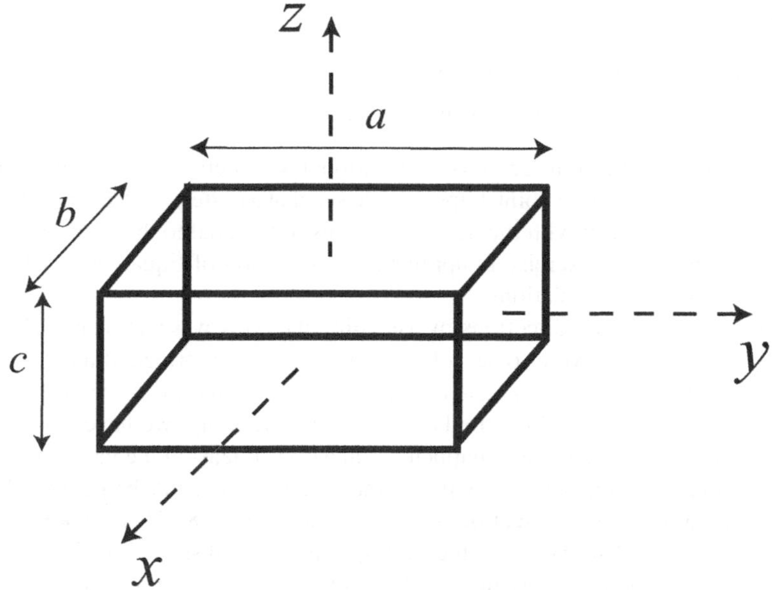

Figure 8.18 Uniform cuboid.

Our next example is a cuboid of uniform mass density ρ, as shown in Figure 8.18. To calculate the inertia tensor we must use the integral form of Equation (8.50), so that, for example,

$$I_{xy} = -\int \rho xy dV \,. \tag{8.71}$$

For each mass element $dm = \rho dV$ at position (x, y, z) there is another one at $(-x, y, z)$, so over the entire body Equation (8.71) integrates to zero. Also, from Equation (8.22) we have that $I_{xx} = Ma^2/12$. The complete inertia tensor is therefore

$$\mathbf{I} = \frac{M}{12}\begin{pmatrix} a^2 & 0 & 0 \\ 0 & b^2 & 0 \\ 0 & 0 & c^2 \end{pmatrix}, \tag{8.72}$$

which I suspect you guessed. Note that since there are three symmetry planes, and any axis perpendicular to a symmetry plane is principal, the choice of origin doesn't matter, though this is unusual. If $a = b = c$ we have a cube, and all three axes are degenerate with $I_{xx} = I_{yy} = I_{zz} = \frac{1}{12}ma^2$, and the body becomes a spherical top. Although we calculated with axes aligned to the body's edges, we could have chosen any orientation about the cube's centre and obtained the same result! From the point of view of rotating bodies, a cube and a sphere are identical: when either body is rotated about an axis passing through its centre, the angular momentum points along the axis, no matter how the axis is oriented.

We can now understand that the contortions we went through in setting up the elementary calculations of moment of inertia in Section 8.4 were really to ensure that the system would be rotating about a principal axis. There are two useful facts that connect a body's symmetry to its principal axes:

1. a rotational symmetry axis is principal;
2. any axis normal to a symmetry plane is principal.

Note that any plane containing an axis of rotational symmetry is a symmetry plane, so applying point 2 in cases where point 1 applies, we see that a system with an axis of rotational symmetry is a symmetric top with the degenerate axes orthogonal to the rotational symmetry axis. Proving 1 and 2 is an exercise in applying the definition of Equation (8.50). They can often be used to simplify calculations.

We found out the principal directions by assuming the body was stationary. The principal directions tell us that *if* we were to set a body rotating about a principal axis, then there is no couple at the axis. The principal axes are all about the *potential* of what would happen if the body were set to rotate. Once the body is actually rotating, we have relinquished the possibility of asking the 'what would happen if' question, at least in the lab frame. However, we can still calculate the inertia tensor, with the inertia tensor elements being time-dependent, as we did for the wobbly bar in Section 8.7, Equations (8.52)–(8.57). By taking a snapshot at a particular time we discover how the principal axes are disposed at that instant. At a later time we discover their new orientation. In other words, we can regard the principal axes as being fixed within a body, and rotating with the body once it is set in motion. In some cases there may be some simplification in choosing the coordinate axes to coincide with the principal axes. The important point to then remember is that once the body actually rotates, this represents a *rotating frame*, which is noninertial. Rotating frames will be discussed in Chapter 9.

Let's pull things together. The rate of change of the angular momentum of a body rotating about a fixed axis is the sum of the torques acting on the body (Equation (8.5)). These torques are of two kinds: one type acts to increase the body's rate of rotation and is directed along the rotation axis. We have also seen that if the rotation axis is not a principal axis, then a couple is required from the axis to maintain the rotation. The direction of the couple at the axis is perpendicular to the axis. We therefore have

$$\frac{d\mathbf{L}}{dt} = \mathbf{G}_{\text{applied}} + \mathbf{G}_{\text{axis}} \,, \tag{8.73}$$

where the first term on the right is the sum of all the elementary torques applied by external forces acting and the second term is the couple at the axis. Since $\mathbf{L} = \mathbf{I}\boldsymbol{\omega}$ we have

$$\mathbf{I}\frac{d\boldsymbol{\omega}}{dt} + \frac{d\mathbf{I}}{dt}\boldsymbol{\omega} = \mathbf{G}_{\text{applied}} + \mathbf{G}_{\text{axis}} \,. \tag{8.74}$$

If we assume that there is no net external toque acting parallel to the axis then both $\mathbf{G}_{\text{applied}}$ and $\mathbf{I}\frac{d\boldsymbol{\omega}}{dt}$ vanish since there is no mechanism to increase the angular velocity. In this case

$$\frac{d\mathbf{I}}{dt}\boldsymbol{\omega} = \mathbf{G}_{\text{axis}} \,, \tag{8.75}$$

which is a useful equation to calculate the couple at the axis. In Section 8.7 we did this for the two-mass wobbly bar to find the axial couple at $t = \pi/2\omega$ (Equation (8.61)). You can now combine Equations (8.54), (8.56) and (8.57) with (8.75) to show that for any time, t

$$\mathbf{G}_{\text{axis}} = \frac{m\omega l^2}{2} \sin\theta\cos\theta\,(\sin\omega t\mathbf{i} - \cos\omega t\mathbf{j}) \ . \tag{8.76}$$

The axial couple lies in the x–y plane and rotates about the z-axis.

8.10 Kinetic Energy of a Body Rotating About a Fixed Axis

The inertia tensor is also necessary when considering the kinetic energy of a rotating body. When the body is rotating about a principal axis we saw in Section 8.5 that the kinetic energy is given by $W = \frac{1}{2}I\omega^2$, where I was called 'the' moment of inertia, though we now understand this to mean the relevant principal moment. In order to generalise to situations where the body is not rotating about a principal axis, it is easiest to calculate the total kinetic energy from first principles.

The kinetic energy of an element of a body rotating about a fixed axis is $\frac{1}{2}m_i\,|\boldsymbol{\omega}\times\mathbf{r}_i|^2$ so that the total kinetic energy is given by

$$T = \sum \frac{1}{2}m_i\,|\boldsymbol{\omega}\times\mathbf{r}_i|^2 \ .$$

From the vector identity[5] $|\mathbf{a}\times\mathbf{b}|^2 = a^2b^2 - (\mathbf{a}\cdot\mathbf{b})^2$ we see that

$$T = \sum \frac{1}{2}m_i\left[r_i^2\omega^2 - (\boldsymbol{\omega}\cdot\mathbf{r}_i)^2\right] \ .$$

By comparing with Equations (8.45) and (8.51) we see that

$$T = \frac{1}{2}\mathbf{L}_{\text{total}}\cdot\boldsymbol{\omega} = \frac{1}{2}\boldsymbol{\omega}\cdot(\mathbf{I}\boldsymbol{\omega}) \ . \tag{8.77}$$

When the angular velocity is constant, the angular momentum may still be time dependent, as we saw previously. However, the only kind of time dependence that can occur is for the angular momentum to precess about the rotation axis. This means that the component of $\mathbf{L}_{\text{total}}$ along the direction of $\boldsymbol{\omega}$ is independent of time, and we see from Equation (8.77) that the kinetic energy is then constant. Somehow, any time dependence of $\mathbf{I}$ is cancelled out in the expression $\frac{1}{2}(\mathbf{I}\boldsymbol{\omega})\cdot\boldsymbol{\omega}$ (essentially through the identity $\sin^2\omega t + \cos^2\omega t = 1$). Of course if there is an external torque acting around the rotation axis, then the system will have an angular acceleration, and T will increase with time. We can use the fact that T is independent of time when there is no external torque along the axis, to make a useful simplification. Recall that the principal axes rotate with the body. However, at any instant, the principal axes provide an alternative coordinate system to the original Cartesian axes. So at a given instant we can write the angular velocity as $\boldsymbol{\omega} = \omega_1\mathbf{e}_1 + \omega_2\mathbf{e}_2 + \omega_3\mathbf{e}_3$, where $(\mathbf{e}_1, \mathbf{e}_2, \mathbf{e}_3)$ are orthonormal vectors aligned with the principal axes. In this coordinate system, the angular momentum is expressed as

$$\mathbf{L} = I_1\omega_1\mathbf{e}_1 + I_2\omega_2\mathbf{e}_2 + I_3\omega_3\mathbf{e}_3 \ , \tag{8.78}$$

and Equation (8.77) then takes the particularly simple form:

$$T = \frac{1}{2}\left(I_1\omega_1^2 + I_2\omega_2^2 + I_3\omega_3^2\right) \ . \tag{8.79}$$

[5] You can prove this simply by writing out the components.

Now it might look as though we are going to have to repeat the calculation of kinetic energy at a later time, since the principal axes will by then have rotated. However, since T is constant we only have to use Equation (8.79) once. As a simple example, imagine rotating the cuboid of Figure 8.18 about its longest diagonal. At the instant when the cuboid is aligned with the cartesian axes as in the figure, we have that

$$\boldsymbol{\omega} = \left[\frac{(a\mathbf{e}_1 + b\mathbf{e}_2 + c\mathbf{e}_3)}{(a^2 + b^2 + c^2)^{1/2}}\right]\omega\ , \tag{8.80}$$

since at this time $\mathbf{e}_1 = \mathbf{i}, \mathbf{e}_2 = \mathbf{j}, \mathbf{e}_3 = \mathbf{k}$. Combining Equations (8.72), (8.79) and (8.80) gives

$$T = \frac{m\omega^2\left(a^4 + b^4 + c^4\right)}{24\left(a^2 + b^2 + c^2\right)}\ , \tag{8.81}$$

or for a cube, $T = m\omega^2 a^2/24$.

8.11 Summary

Section 8.1. A **rigid body** is a system in which the forces act so as to maintain the distance between its constituent particles.

Section 8.2. The **total torque** acting on a system of particles equals the rate of change of its **total angular momentum**

$$\mathbf{G}_{\text{total}} = \frac{d\mathbf{L}_{\text{total}}}{dt}\ , \tag{8.5}$$

where

$$\mathbf{G}_{\text{total}} = \mathbf{r}_1 \times \mathbf{F}_1^{\text{ext}} + \mathbf{r}_2 \times \mathbf{F}_2^{\text{ext}}\ , \tag{8.3}$$

and

$$\mathbf{L}_{\text{total}} = \mathbf{r}_1 \times \mathbf{p}_1 + \mathbf{r}_2 \times \mathbf{p}_2 = \mathbf{L}_1 + \mathbf{L}_2 + \cdots\ . \tag{8.4}$$

Section 8.3. The centre of mass of a continuous distribution of matter is given by

$$\mathbf{R} = \frac{\int \mathbf{r}\rho(\mathbf{r})dV}{\int \rho(\mathbf{r})dV} = \frac{\int \mathbf{r}\rho(\mathbf{r})dV}{M}\ . \tag{8.10}$$

Section 8.4. For a uniform body possessing a plane of symmetry rotating with angular speed ω, about an axis orthogonal to the symmetry plane

$$L_{\text{total}} = \int \omega s^2 \rho dV = \omega I\ , \tag{8.15}$$

where $I \equiv \int \rho s^2 dV$ is the body's **moment of inertia**. If I_{CM} is the moment of inertia about a parallel axis passing through the body's centre of mass then the parallel axis theorem asserts

$$I = MD_{\text{CM}}^2 + I_{\text{CM}}\ , \tag{8.19}$$

where D_{CM} is the distance between the two axes.

Section 8.5. The angular work–energy theorem

$$W = \int_{\omega_1}^{\omega_2} I\omega d\omega = \frac{1}{2}I\omega_1^2 - \frac{1}{2}I\omega_2^2 \, . \qquad (8.28)$$

Section 8.7. For a general body rotating about a fixed axis

$$\mathbf{L}_{\text{total}} = \mathbf{I}\boldsymbol{\omega} \, , \qquad (8.51)$$

where $\mathbf{I}$ is the **inertia tensor**, whose diagonal elements are given by $I_{xx} = \sum_i m_i(y_i^2 + z_i^2)$ etc. and whose off-diagonal elements are given by $I_{xy} = -\sum_i m_i x_i y_i$ etc.

Section 8.8. On account of its symmetry ($I_{xy} = I_{yx}$ etc.) the inertia tensor can always be diagonalised (see Appendix, Section A.7), meaning that there exists a coordinate system (the **principal coordinates**) for which the inertia tensor can be expressed as

$$\mathbf{I} = \begin{bmatrix} I_1 & 0 & 0 \\ 0 & I_2 & 0 \\ 0 & 0 & I_3 \end{bmatrix} , \qquad (8.63)$$

where $I_{1,2,3}$ are called the **principal moments of inertia**. They are found by solving for the eigenvalues λ in

$$\det |\mathbf{I} - \lambda \mathbf{Id}| = 0 \, , \qquad (8.65)$$

where $\mathbf{I}$ is expressed in the original cartesian coordinate system. The corresponding eigenvectors, found using the method described in the Appendix (Section A.6) identify the **principal axes** of the system.

Section 8.9. When no external torque acts parallel to the rotation axis, then the couple at the axis is given by

$$\frac{d\mathbf{I}}{dt}\boldsymbol{\omega} = \mathbf{G}_{\text{axis}} \, . \qquad (8.75)$$

Section 8.10. The kinetic energy of a body rotating about a fixed axis is given by

$$T = \frac{1}{2}\mathbf{L}_{\text{total}} \cdot \boldsymbol{\omega} = \frac{1}{2}\boldsymbol{\omega} \cdot (\mathbf{I}\boldsymbol{\omega}) \, . \qquad (8.77)$$

This is often simplified by using coordinates referred to the principal axes, in which case

$$T = \frac{1}{2}\left(I_1\omega_1^2 + I_2\omega_2^2 + I_3\omega_3^2\right) \, . \qquad (8.79)$$

8.12 Problems

8.1 Calculate the moment of inertia of the following symmetric bodies:

(a) a cube of side L rotating about one edge;

(b) a sphere of radius L rotating about its centre of mass;
(c) a sphere of radius L rotating about a point on its surface;
(d) a solid cylinder rotating about its axis;
(e) a hollow cylinder rotating about its axis.

8.2 A nitrogen molecule, N_2, can be regarded as a rigid system consisting of two point masses separated by 1.1×10^{-10} m. If the angular momentum of rotation about an axis through the centre of mass and perpendicular to the line of centres is one quantum of angular momentum ($= h/2\pi$), calculate the frequency of the rotation. What is the wavelength of electromagnetic radiation which could excite this rotational motion? ($h/2\pi = 10^{-34}$ J s; atomic weight of nitrogen = 14; Avagadro's number $N_0 = 6 \times 10^{26}$ per kg mole.)

8.3 A cylinder of radius R and mass m rolls without slipping down a plane inclined at an angle θ to the horizontal (see Figure 8.19). In time dt the angle through which the

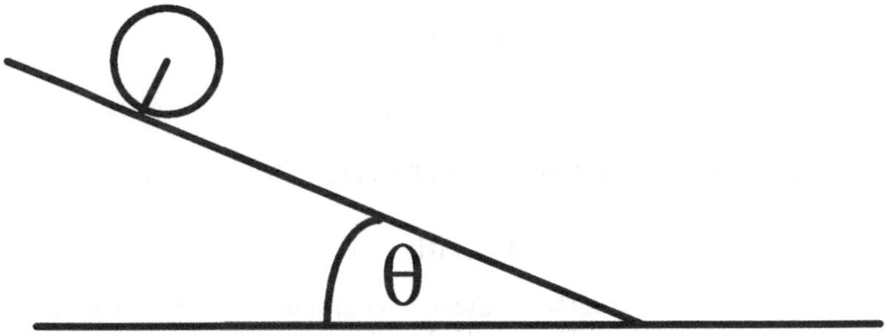

Figure 8.19 Figure for Problem 8.3.

cylinder turns is $d\phi$. Show that:

(a) the distance moved by the cylinder is $ds = Rd\phi$;
(b) the speed of the cylinder is $v = R\dot{\phi} = R\omega$;
(c) the work done by gravity when the cylinder rolls down a length s of the plane is $mgs \sin \theta$.
(d) Hence by applying the work–energy theorem show that

$$mgs \sin \theta = \frac{1}{2} I\omega^2 + \frac{1}{2} mv^2 ,$$

where I is the moment of inertia of the cylinder about its central axis (hint: the work acts to increase both the linear and angular kinetic energy).
(e) If $I = \alpha m R^2$, where α is a constant, show that the time taken for the cylinder to reach the bottom, and the speed when it arrives, are respectively given by

$$t = \left[\frac{2(1+\alpha)s}{g \sin \theta} \right]^{1/2} \quad \text{and} \quad v = \left[\frac{2gs \sin \theta}{1+\alpha} \right]^{1/2} .$$

(f) A hollow cylinder and a uniform solid cylinder have the same mass and are released from rest. Which will reach the bottom of the inclined plane first? Calculate the ratio of their respective arrival times at the bottom.

8.4 In the situation illustrated in Figure 8.20 the axis is fixed horizontally and the other end of the rod is supported so that the rod rests horizontally. What is the force at the axis? The support at the far end of the rod is removed. By considering the acceleration of the centre of mass, show that the force at the axis, immediately after removal of the support, is given by $F = Mg/4$.

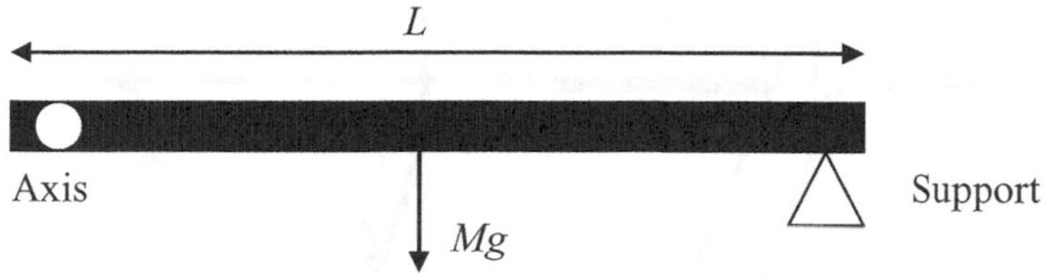

Figure 8.20 Figure for Problem 8.4.

8.5 Find the principal axes of inertia and the corresponding moments of inertia of the following systems rotating about their centres of mass (choose the origin to be at the centre of mass).

(a) two masses m_1 and m_2 separated by a light rod of length l;

(b) a water molecule, considered as a mass m_1 connected by light rods of length l to masses m_2, the angular separation of the rods being 2θ;

(c) an ammonia molecule modelled as a regular tetrahedron of side l with a point mass m_1 at one vertex, and three point masses m_2 at the other vertices.

Indicate which if any of these is a spherical or symmetric top.

8.6 (a) For a system of three identical mass points located at $(a, 0, 0)$, $(0, a, 2a)$ and $(0, 2a, a)$, find the principal moments of inertia and a set of principal axes.

(b) Calculate the rotational kinetic energy when the system rotates about the $(1, 1, 1)$ axis with angular velocity 10 rad s^{-1}.

(c) A fourth identical mass is now located at $\left(2, \sqrt{3}, \sqrt{3}\right) a$. Given that the unit vector $\mathbf{e} = \left(0, 1/\sqrt{2}, -1/\sqrt{2}\right)$ identifies a principal axis for the new system, calculate the new principal moment associated with this axis.

8.7 (a) Draw a diagram indicating a choice of principal axes for a uniform circular disc of mass M and radius a. Show that the principal moment about an axis perpendicular to the disc through its centre is given by $I_1 = \frac{1}{2}Ma^2$, whilst the other two principal moments are determined from

$$I_2 = I_3 = \frac{2Ma^2}{\pi} \int_{-\frac{\pi}{2}}^{\frac{\pi}{2}} \sin^2\theta \cos^2\theta d\theta = \frac{Ma^2}{4} .$$

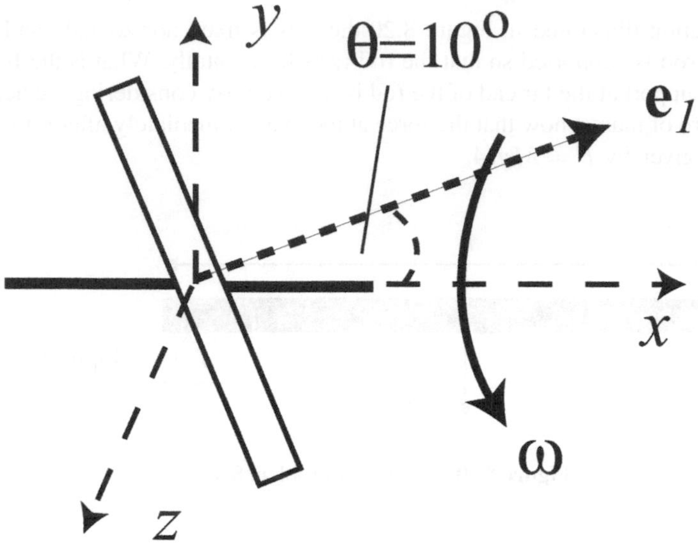

Figure 8.21 Figure for Problem 8.7

(b) A disc of mass $M = 3$ kg and radius $a = 0.2$ m is constrained to spin at 50 rev s^{-1} about a light axle placed along the x-axis as shown in Figure 8.21. Note that the principal axis identified by the unit vector $\mathbf{e}_1$ is constrained to precess about the x-axis at an angle of $\theta = 10^\circ$.

In the indicated Cartesian frame the relevant components of the inertia tensor are given by

$$I_{xx} = I_1 \cos^2\theta + I_2 \sin^2\theta \,,$$

$$I_{yx} = \frac{1}{2}(I_1 - I_2)\cos\omega t \sin 2\theta \,,$$

$$I_{zx} = \frac{1}{2}(I_1 - I_2)\sin\omega t \sin 2\theta \,.$$

Calculate the system kinetic energy and the magnitude of the couple on the axle.

8.8 Figure 8.22 shows the rear view of an airborne helicopter that is inclined at 30° to the vertical. The craft is in vertical equilibrium. The four rotor blades, each of mass 50 kg and length 4 m, rotate anticlockwise when viewed from above. The remaining mass of the craft is 4000 kg and its centre of mass, C, is located 2 m below the rotor blades as shown.

(a) Calculate the principal moment of inertia of the rotor blades about the principal axis AB.

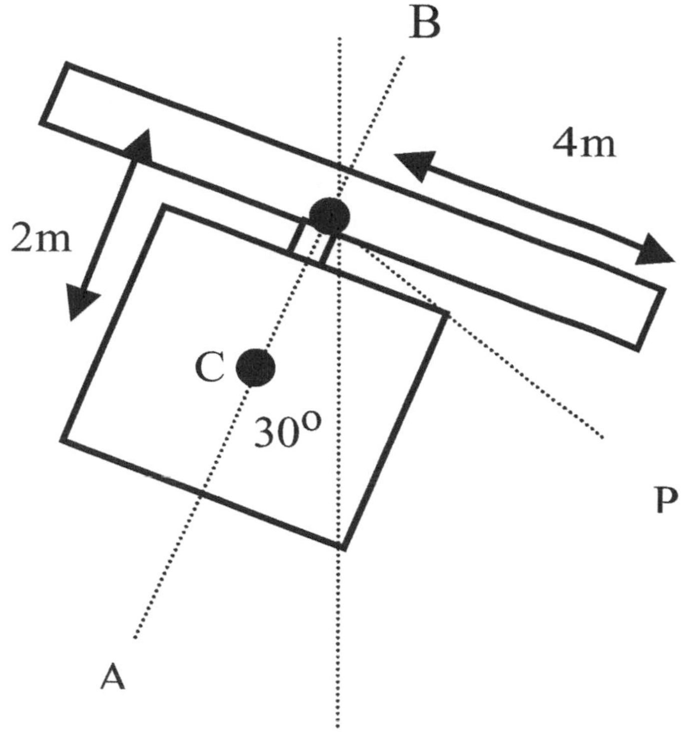

Figure 8.22 Figure for Problem 8.8.

(b) The lift provided by the rotor blades is proportional to the angular rotor speed. Given that the rotor blades must rotate at 80 rev s^{-1} for vertical take off, calculate the angular rotor speed when the craft is inclined as shown.

(c) Draw a diagram indicating the forces acting on the craft and the direction of the angular momentum of the blades about the axis. Calculate the net horizontal force.

(d) By considering the torque exerted by the bulk of the craft about point P show that the angular momentum vector in (c) precesses at 0.13 rad s^{-1}. Indicate on your diagram the sense of precession.

(e) Suppose now that the helicopter is circling horizontally with angular speed ω and radius R. Explain briefly why it is easier to circle in one sense than the other, stating the easier sense. If, when circling in the easier sense, the precession angular frequency is matched to ω, calculate R.

9

Rotating Frames

9.1 Introduction

In Chapter 1 we looked closely at what constituted an inertial frame, noting that this was an idealisation approached in practice by successive approximation. We also showed that any attempt to use Newton's laws in accelerating frames must be accompanied by the artificial introduction of so-called inertial forces which cancel off the effects of the frame's acceleration. We now turn to a particular case of an accelerating frame, namely a frame which is *rotating*. This is evidently a relevant case to consider for any terrestrial experiments since the Earth rotates at $\omega = 2\pi/(24 \times 60 \times 60) = 7.3 \times 10^{-5}$ rad s^{-1}.

The notion of a rotating frame, and the attempt to use Newton's laws therein, needs careful consideration. For an inertial frame, apart from the difficulties of removing extraneous matter, there is no intrinsic limitation on the frame's extent. Even in the absence of matter, however, a rotating frame[1] must be localised if the discussion is going to be kept within the framework of Newtonian physics. A particle at rest in the rotating frame, distant R from the origin, will move at the tangential speed ωR with respect to an inertial frame. Relativistic effects will therefore be significant if $R \sim c/\omega$.[2] This, however, is not usually a major restriction for macroscopic bodies. For example, relativistic effects become important in the rotating frame associated with the Earth at a distance of $\sim 4 \times 10^{12}$ m. Although this is much less than the distance to the nearest star, a body observable from the Earth at this distance would have to possess enormous kinetic energy to appear stationary to an Earthbound observer. The difficulty is potentially more acute for analysing electrons orbiting nuclei and moving at relativistic speeds, though of course classical mechanics is mostly an inappropriate description at this scale.[3] We will be dealing here with macroscopic bodies, for which the general principles will be similar to those considered in Chapter 1 for a linearly accelerating frame, although the details will need to be worked through carefully.

[1] The concept of a rotating frame in the absence of extraneous matter is controversial. Ernst Mach, in his 'Principle of Inertia' claimed that the removal of extraneous matter extinguished any standard against which a frame could be said to be rotating. In this view, removing all extraneous matter surrounding the Earth would cause the plane through which a Foucault pendulum swings to cease rotating.

[2] In fact when linearly accelerating frames are analysed using relativity there turns out to be a restriction associated with their size also. The frame's extent in the direction of acceleration (a) must be much less than c^2/a .

[3] So-called Thomas precession of the electron spin axis is an important relativistic effect due to the electron's orbital motion which can be analysed without quantum mechanics.

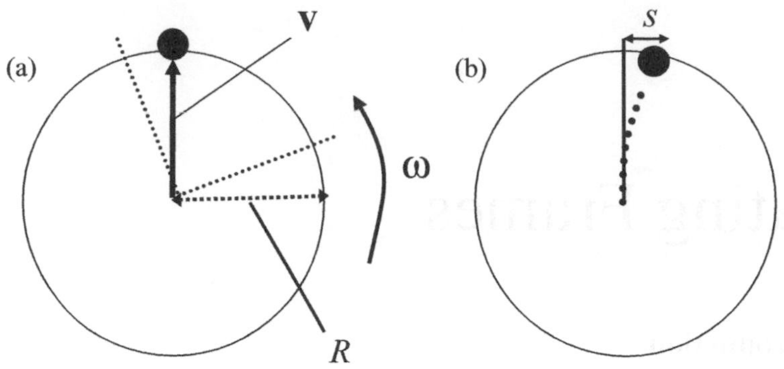

Figure 9.1 Radial motion of a ball on a roundabout: (a) from an inertial observer's viewpoint, (b) from the viewpoint of someone at rest with respect to the roundabout.

9.2 Experiments on Roundabouts

To get our bearings, consider the simple experiment illustrated in Figure 9.1. I take my son, Thomas, to the local playground and ask him to sit on the middle of a roundabout (radius R, rotating at ω) and to project a ball (of mass m) towards the edge (with speed v'). I don't like getting giddy so I watch from the side. From my inertial viewpoint, the ball is not subjected to any force and follows the straight line shown in Figure 9.1(a). Thomas, however, rotating with the roundabout, sees the ball follow the curved trajectory of Figure 9.1(b). Even if he projects the ball quickly (so that $v' \gg \omega R$, but still of course $\ll c$) it will nevertheless appear to him to be deflected sideways by a small distance, s. To this sideways movement he can ascribe an apparent acceleration, a' (the prime is to remind us that this is a quantity measured in a noninertial frame). Since the rim of the roundabout moves through a distance ωRt in the time t it takes the ball to reach the rim, he calculates

$$s = \frac{1}{2}a't^2 \approx \omega Rt \, . \tag{9.1}$$

Given that the sideways deflection is small, $t \approx R/v'$ so that

$$a' = 2\omega v' \, . \tag{9.2}$$

He can even assign a force to this acceleration, $F' = ma' = 2m\omega v'$, and give it a name, the **Coriolis force**. We emphasise that the ball does not 'feel' any force. In my (inertial) frame the motion is a straight line (Newton's first law). The roundabout (and Thomas) 'feel' (radial) forces to keep *them* rotating, and relative to them the ball *appears* to accelerate. When analysing motion in this rotating frame, if Newton's laws are going to be applied, then a force must be invented that gives rise to this *apparent* acceleration. For the counter-clockwise rotation of Figure 9.1 the force and the subsequent deflection are to the right.

Now consider tangential motion (just as in the case of forces considered in Chapter 7, it is convenient to consider radial and tangential motion separately.). Imagine that the roundabout has a wall around its rim, like a roulette wheel, so that when I ask Thomas to sit at the rim and project the ball tangentially it follows the circumferential path shown in Figure 9.2. The ball is constrained to move in a circle so that according to me, if its tangential speed is v, then its acceleration is $(-v^2/R)$, the minus sign indicating that it is directed radially inwards.[4] The force supplied by the rim of the roundabout compelling the ball to move in a circle is

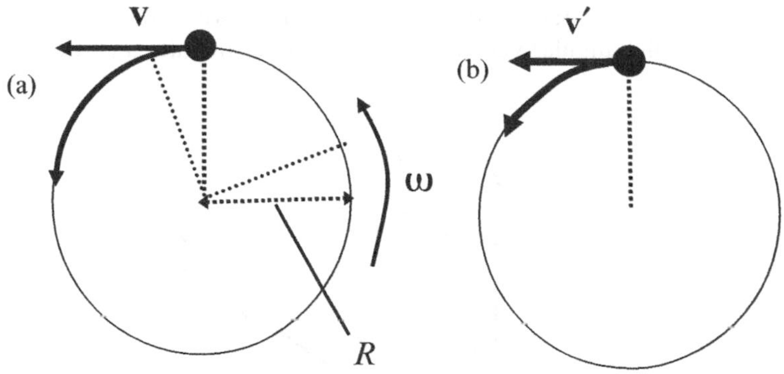

Figure 9.2 Tangential motion of a ball on a roundabout: (a) from an inertial observer's viewpoint, (b) from the viewpoint of someone at rest with respect to the roundabout.

therefore $(-mv^2/R)$. Now Thomas, moving at the rim with speed ωR, assigns a tangential speed v' to the ball so that

$$v = v' + \omega R\,. \tag{9.3}$$

The acceleration I observe can therefore be written as

$$a = -\frac{v^2}{R} = -\frac{(v' + \omega R)^2}{R}\,,$$

or

$$a = -\frac{(v')^2}{R} - 2v'\omega - \omega^2 R\,. \tag{9.4}$$

The first term on the right is the acceleration Thomas would *like* to assign as being that required to keep the ball moving in a circle in *his* frame, a'. However, for the picture to be correct, he must add on two further terms so that

$$a' = a + 2\omega v' + \omega^2 R\,. \tag{9.5}$$

The second term on the right depends on the velocity of the particle and is just like the one encountered in Equation (9.2). It is again a Coriolis acceleration, dependent upon the ball's velocity. For the velocity direction shown in Figure 9.2 the Coriolis acceleration is radially outwards, but if the ball were projected around the rim in a clockwise sense, then this acceleration would be radially inwards. The third term is velocity-independent but does depend on how far away the particle is from the rotation axis. It is always directed radially outwards, and is referred to as the **centrifugal acceleration**.

[4]The position vector of a particle travelling in a circle of radius R, with angular speed $\omega_0 = v/R$, is $\mathbf{r} = R\cos\omega_0 t\mathbf{i} + R\sin\omega_0 t\mathbf{j} = R\hat{\mathbf{r}}$, so that its acceleration is $\ddot{\mathbf{r}} = -\omega_0^2\mathbf{r} = -\omega_0^2 R\hat{\mathbf{r}} = (-v^2/R)\,\hat{\mathbf{r}}$.

9.3 General Prescription for Rotating Frames

Of course what we really want is a general prescription for relating the acceleration of a particle between inertial and rotating frames, given, as is generally the case, the particle's velocity in the rotating frame. Equations (9.2) and (9.5) achieve this for the special cases of a particle moving radially and tangentially in two dimensions.

Figure 9.3 shows a rotating frame and a stationary frame in which the z-axes of both frames are coaligned with $\boldsymbol{\omega}$, the angular velocity vector. Consider initially a particle at position $\mathbf{r}$,

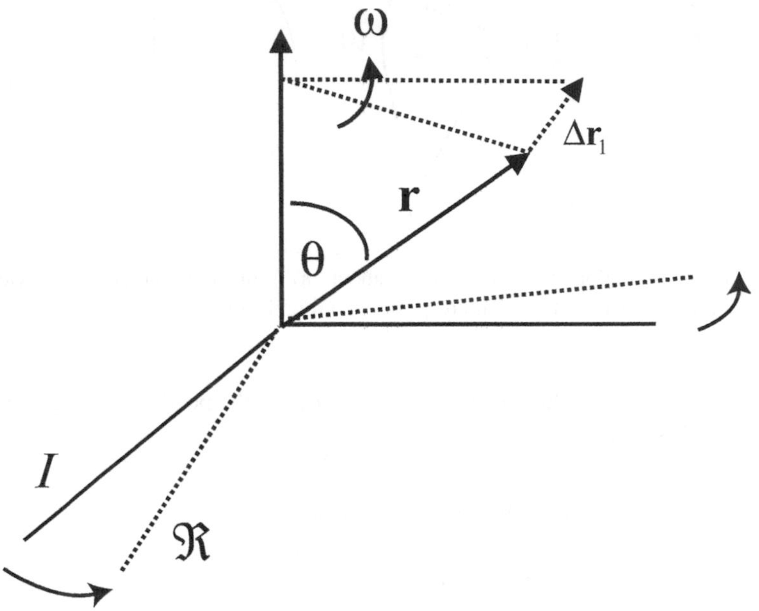

Figure 9.3 A rotating frame and an inertial frame, with a particle at rest in the rotating frame.

at rest in the rotating frame. In time dt the rotating frame rotates through an angle ωdt so that the change in the particle's position in the inertial frame is given by

$$|dr_1| = \omega r \sin\theta \; dt \,, \tag{9.6}$$

where the subscript 1 on $d\mathbf{r}$ is to indicate that the variation is due solely to the frame rotation (we will introduce a second type of variation in a moment). The direction of $d\mathbf{r}_1$is parallel to $\boldsymbol{\omega} \times \mathbf{r}$, so

$$\frac{d\mathbf{r}_1}{dt} = \boldsymbol{\omega} \times \mathbf{r} \,. \tag{9.7}$$

Now if the particle is not stationary in the rotating frame, but moves an amount $d\mathbf{r}_2$ during dt then the situation is as shown in Figure 9.4. The total displacement during dt is therefore

$$d\mathbf{r} = d\mathbf{r}_1 + d\mathbf{r}_2 = d\mathbf{r}_2 + (\boldsymbol{\omega} \times \mathbf{r}) \, dt \,, \tag{9.8}$$

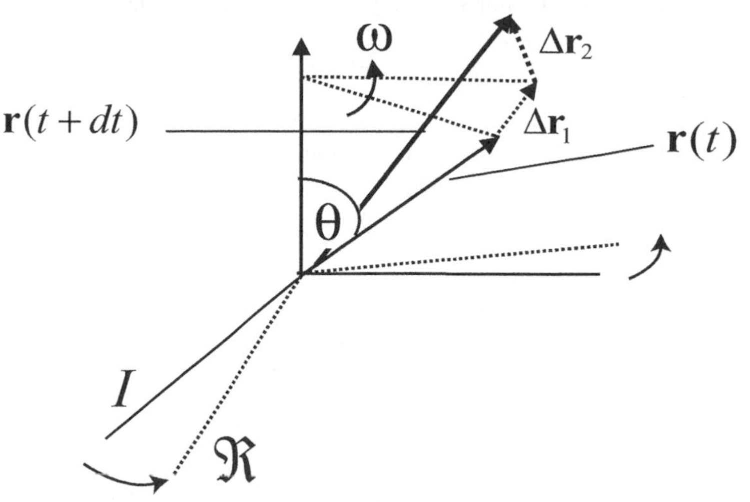

Figure 9.4 A rotating frame and an inertial frame, with a particle moving in the rotating frame.

or

$$\frac{d\mathbf{r}}{dt} = \frac{d\mathbf{r}_2}{dt} + \boldsymbol{\omega} \times \mathbf{r} \,. \tag{9.9}$$

There are thus two sorts of variation of position with respect to time, the *intrinsic* variation due to the particle's movement in the rotating frame ($= d\mathbf{r}_2/dt$), and the *total* variation observed in the inertial frame which includes a contribution due to the frame rotation ($= d\mathbf{r}/dt$). We need to distinguish between the two sorts of variation, so we will subscript with $\Re$ time derivatives in the rotating frame, and with I time derivatives in the inertial frame:

$$\left(\frac{d}{dt}\right)_I \mathbf{r} = \left(\frac{d}{dt}\right)_{\Re} \mathbf{r} + \boldsymbol{\omega} \times \mathbf{r} \,, \tag{9.10}$$

or

$$\mathbf{v}_I = \mathbf{v}_{\Re} + \boldsymbol{\omega} \times \mathbf{r} \,, \tag{9.11}$$

where $\mathbf{v}_I$ and $\mathbf{v}_{\Re}$ represent the particle's velocity observed in the inertial and the rotating frame respectively.

Now the same analysis can be applied to *any* vector, $\mathbf{A}$, not just the position vector, so that in general

$$\left(\frac{d}{dt}\right)_I \mathbf{A} = \left(\frac{d}{dt}\right)_{\Re} \mathbf{A} + \boldsymbol{\omega} \times \mathbf{A} \,. \tag{9.12}$$

Applying this prescription to the inertial frame velocity of the particle, that is setting $\mathbf{A} = (d/dt)_I\, \mathbf{r} = \mathbf{v}_I$, we obtain

$$\left(\frac{d}{dt}\right)_I \mathbf{v}_I = \left(\frac{d}{dt}\right)_{\Re} \mathbf{v}_I + \boldsymbol{\omega} \times \mathbf{v}_I \,. \tag{9.13}$$

The right-hand side of Equation (9.13) now contains a mixture of rotating frame and inertial frame quantities. To express the right-hand side entirely in terms of rotating frame quantities we can substitute for $\mathbf{v}_I$ in terms of $\mathbf{v}_\Re$ using Equation (9.11) so that

$$\left(\frac{d}{dt}\right)_I \mathbf{v}_I = \left(\frac{d}{dt}\right)_\Re (\mathbf{v}_\Re + \boldsymbol{\omega} \times \mathbf{r}) + \boldsymbol{\omega} \times (\mathbf{v}_\Re + \boldsymbol{\omega} \times \mathbf{r}) \ . \tag{9.14}$$

The left-hand side is just the acceleration of the particle measured from the inertial frame, $\mathbf{a}_I$. Assuming the frame angular velocity is constant (it is complicated enough already!), the right-hand side can be developed to yield

$$\mathbf{a}_I = \mathbf{a}_\Re + 2\omega \times \mathbf{v}_\Re + \boldsymbol{\omega} \times (\boldsymbol{\omega} \times \mathbf{r}) \ , \tag{9.15}$$

where $\mathbf{a}_\Re$ is the acceleration of the particle according to the rotating observer.[5] Typically $\mathbf{a}_I = \mathbf{g}$, the acceleration due to gravity at the Earth's surface, and $\mathbf{v}_\Re$ is the specified velocity in the Earth's rotating frame, so that the most useful form for Equation (9.15) is to express the unknown $\mathbf{a}_\Re$ in terms of the remaining quantities as

$$\mathbf{a}_\Re = \mathbf{a}_I - 2\omega \times \mathbf{v}_\Re - \boldsymbol{\omega} \times (\boldsymbol{\omega} \times \mathbf{r}) \ . \tag{9.16}$$

This is the generalisation of Equations (9.2) and (9.5) that we sought. The second and third term on the right are the Coriolis and centrifugal accelerations respectively. The magnitude of the centrifugal term is $\omega^2 r \sin\theta$ (where θ is the angle between $\boldsymbol{\omega}$ and $\mathbf{r}$) and its direction is away from the rotation axis.

9.4 The Centrifugal Term

Equation (9.16) is sufficient to solve any problem posed in a rotating frame. The chief difficulty is generally to get the directions of the Coriolis and centrifugal terms correct. In order to obtain a clearer picture of what is going on it is useful to view the Earth-bound rotating frame from a point in the solar system removed from the Earth (see Figure 9.5). Attach orthogonal unit vectors $\mathbf{i}, \mathbf{j}$ and $\mathbf{k}$ to a point $\mathbf{r}$ at the Earth's surface at latitude λ ($\mathbf{i}$ is due East, $\mathbf{j}$ is due North and $\mathbf{k}$ is 'up'). The latitude, λ, is related to the *colatitude*, θ, by $\lambda = 90^o - \theta$. Now $(\boldsymbol{\omega} \times \mathbf{r})$ is due east, so $-\boldsymbol{\omega} \times (\boldsymbol{\omega} \times \mathbf{r})$ is directed away from the Earth's axis as shown. Projecting the centrifugal acceleration vector onto the $\mathbf{i}$, $\mathbf{j}$, $\mathbf{k}$-axes gives

$$-\boldsymbol{\omega} \times (\boldsymbol{\omega} \times \mathbf{r}) = \omega^2 R_\mathrm{E} \cos\lambda \,(-\sin\lambda \mathbf{j} + \cos\lambda \mathbf{k}) \ , \tag{9.17}$$

where R_E is the radius of the Earth ($= 6.37 \times 10^6$ m). In the above equation the second term on the right is a small correction to the gravitational acceleration whose size can be estimated from $\omega^2 R_E \approx 3.4 \times 10^{-2}$ ms^{-2}, that is about one three hundredth of g. The first term causes a plumb-line to hang slightly off from the true vertical, the angular extent of the deviation being given approximately by

$$\alpha \approx \frac{\omega^2 R_E}{g} \sin\lambda \cos\lambda \ . \tag{9.18}$$

In fact, the Earth is not perfectly spherical and, precisely because of centrifugal effects, the diameter at the equator is greater than at the poles. This oblateness causes a plumb-line to *appear* to hang vertically, because the normal to the tangent plane at the Earth's surface does not point exactly to the centre of the Earth (effectively $\mathbf{k}$ is not precisely parallel to $\mathbf{r}$). For this reason the centrifugal term does not impart any deviations within the plane of the Earth's surface. However, the Coriolis term can give rise to such deviations, which we now consider.

[5] The brackets in the vector triple product are essential.

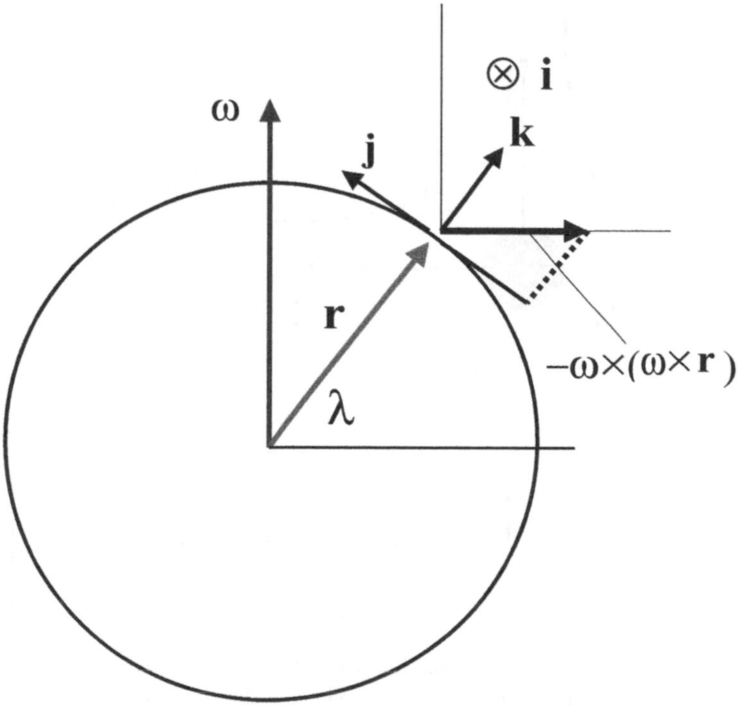

Figure 9.5 Earthbound coordinates at latitude λ.

9.5 The Coriolis Term

Now we consider the direction of deviations occurring in the **i**,**j**-plane due say to projecting a ball along a snooker table lying parallel to the Earth's surface at some northern latitude. This is not the only way that Coriolis effects can occur, but it is the context in which they are usually observed. In terms of **i**,**j**,**k**, the angular velocity $\boldsymbol{\omega}$ is given by

$$\boldsymbol{\omega} = \omega\left(\cos\lambda\mathbf{j} + \sin\lambda\mathbf{k}\right) \ . \tag{9.19}$$

The ball's velocity along the table, **v** (understood to be in the rotating frame, so we now drop the $\Re$ subscript), has only **i** and **j** components. Imagine that unlike a real snooker table there is a hole right in the centre of the table into which we attempt to cue a ball from various directions as shown in Figure 9.6. The direction of the Coriolis acceleration is in the direction of $\mathbf{v} \times \boldsymbol{\omega}$. Now the cross product of **v** with the first term of Equation (9.19) lies in the $\mathbf{k}$ direction and so does not give rise to any planar deviations, but is another small correction to gravity. Planar deviations are due to taking the cross product of **v** with the second term of Equation (9.19) leading to the directions shown in Figure 9.6. An exercise is to show that providing **v** remains in the **i**,**j**-plane, the magnitude of the Coriolis acceleration in the plane is independent of the direction of **v** and given by $2\omega v \sin\lambda$. The figure can also be interpreted as the velocity flow of bath water approaching the plug hole and shows that in the northern hemisphere the water deviates to the right resulting in an anticlockwise circulation. However,

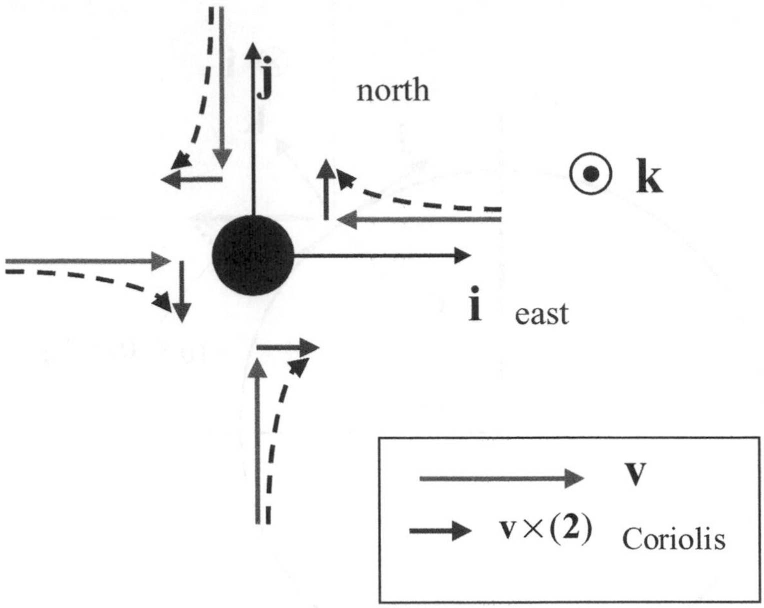

Figure 9.6 Coriolis effects in the plane of the Earth's surface. A ball is cued towards a hole in the centre of a snooker table from various directions and deviates as shown due to the Coriolis effect.

because the magnitude of the Coriolis acceleration is so small, the actual circulation of bath water is probably determined more by other extraneous factors than by the Coriolis force. Of more practical significance is the observation that winds, notionally travelling towards the centre of a depression (since their flow is 'downhill' with respect to the isobars), are apparently pulled rightwards, resulting again in an anticlockwise circulation.

Are snooker shots really affected by the Coriolis acceleration? Figure 9.7 shows a shot covering the full length of a table of length L. The travel time, t, is then $t = L/v$ so that the sideways deflection is given by

$$s = \frac{1}{2}(2\omega v \sin\lambda)\left(\frac{L}{v}\right)^2 = \frac{\omega L^2 \sin\lambda}{v} . \tag{9.20}$$

The effect is most pronounced at either pole(!), is greater the longer the table and is inversely proportional to the speed, v. The last point is a little surprising since the Coriolis acceleration is proportional to v. However, Equation (9.20) shows that this is more than compensated for by the fact that the travel time is inversely proportional to v^2. The deviation is therefore greatest for a long slow pot. For snooker played in London (latitude $\lambda = 52^o$) and for $v = 0.5$ ms^{-1} we obtain $2\omega v \sin\lambda = 5.7 \times 10^{-5}$ ms^{-2}. Setting $L = 2$m we obtain

$$s = \frac{1}{2} \times \left(5.7 \times 10^{-5}\right) \times \left(\frac{2}{0.5}\right)^2 \approx 0.5 \times 10^{-3}\ \mathrm{m} . \tag{9.21}$$

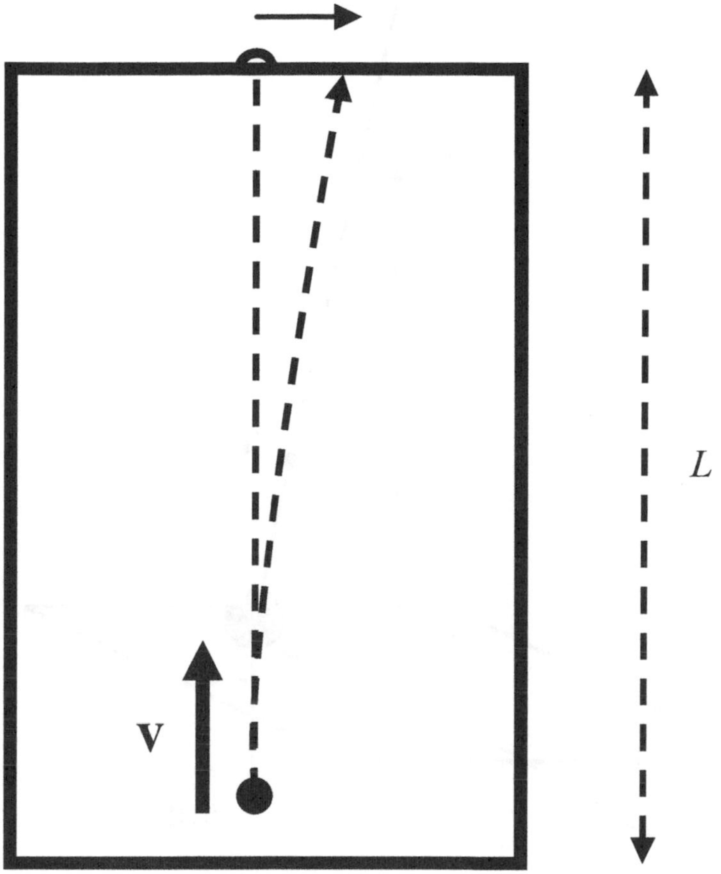

Figure 9.7 Coriolis effects on snooker shots.

The deflection to the right is about half a millimetre. Although this is a rather small deflection, it *ought* to be detectable by a professional player, especially if this lateral deviation of the cue ball induces an *angular* error on the ball to be potted.[6]

A context in which the Coriolis effect definitely does play a significant role is explored in the next section and in problems at the end of this chapter.

9.6 The Foucault Pendulum

A Foucault pendulum consists of a mass suspended from a cord of length l suspended vertically (the z direction) at latitude λ, the rotation plane being free to move in the x-y plane (see Figure 9.8). You can see that we are again using the Earthbound coordinate system

[6]In preparing the first edition of this book I contacted the former world snooker champion, Steve Davis, who said that throughout his career he had indeed noted a small rightwards deviation. However, the conclusive test of the deviation being to the left in the southern hemisphere awaits confirmation!

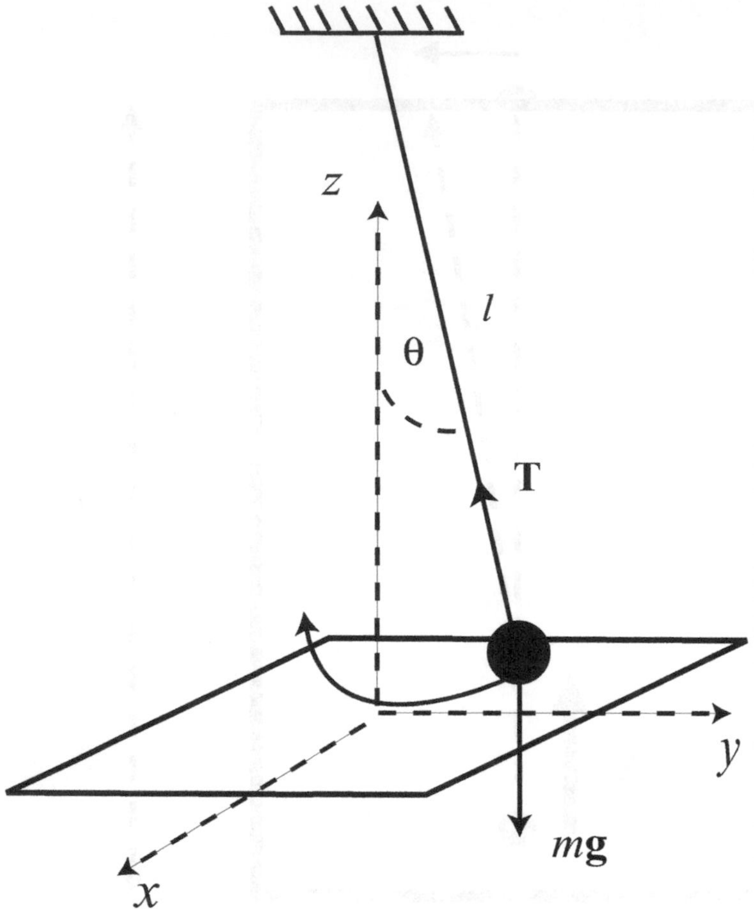

Figure 9.8 The Foucault pendulum.

of Figure 9.5. The tension

$$\mathbf{T} = T\left[-\left(\frac{x}{l}\right)\mathbf{i} - \left(\frac{y}{l}\right)\mathbf{j} + \cos\theta\mathbf{j}\right] . \tag{9.22}$$

Equating the z-component of this expression to the weight mg we see that for small angles ($\cos\theta \approx 1$), $T \approx mg$. The acceleration on the mass in the x-y plane due to the tension in the string is therefore

$$\mathbf{a}_{\mathrm{T}} = -\frac{g}{l}(x\mathbf{i} + y\mathbf{j}) . \tag{9.23}$$

Since the laboratory frame is rotating, we must also include a term to take account of the Coriolis effect. Combining Equation (9.19) for $\boldsymbol{\omega}$ with $\mathbf{v}_{\Re} = \dot{x}\mathbf{i} + \dot{y}\mathbf{j}$, the Coriolis acceleration in the x-y plane is given by

$$\mathbf{a}_{\mathrm{C}} = -2(\boldsymbol{\omega} \times \mathbf{v}_{\Re}) = 2\omega \sin\lambda\,(\dot{y}\mathbf{i} - \dot{x}\mathbf{j}) . \tag{9.24}$$

As explained at the end of Section 9.4 centrifugal effects can be ignored in this problem. So the equations for the total accelerations in the x and y directions are

$$\ddot{x} - 2\Omega\dot{y} + \omega_0^2 x = 0 ,$$
$$\ddot{y} + 2\Omega\dot{x} + \omega_0^2 y = 0 ,$$

where $\omega_0^2 = g/l$ and $\Omega = \omega \sin\lambda$. A neat way to solve this system is to introduce, purely as a mathematical device, the complex number $z = x + iy$. Then the two equations can be written as the single equation

$$\ddot{z} + 2i\Omega\dot{z} + \omega_0^2 z = 0 . \tag{9.25}$$

This now looks very similar to the equation for a damped oscillator which we considered in Section 3.10 (cf. Equation (3.46)). The differences are that here there is no forcing term on the right, and the 'damping' coefficient is imaginary. However we can still use the same method to find a solution, that is by proposing a solution of the form $z = z_0 \exp i\mu t$, where μ is some number to be determined. Substituting this in Equation (9.25) yields the quadratic equation

$$\mu^2 + 2\Omega\mu - \omega_0^2 = 0 , \tag{9.26}$$

for which the solutions are $\mu = -\Omega \pm \left(\Omega^2 + \omega_0^2\right)^{1/2}$. The general solution is therefore

$$z(t) = \left[z_+ e^{i\omega' t} + z_- e^{-i\omega' t}\right] e^{-i\Omega t} , \tag{9.27}$$

where $\omega' = \left(\Omega^2 + \omega_0^2\right)^{1/2}$, and $z_\pm$ are complex constants determined by the initial conditions. If we assume the pendulum is released from rest at $x = a, y = 0$, then

$$z(0) = z_+ + z_- = a , \tag{9.28}$$
$$\dot{z}(0) = z_+(\omega' - \Omega) - z_-(\omega' + \Omega) = 0 , \tag{9.29}$$

and therefore

$$z_\pm = \left(\frac{\omega' \pm \Omega}{2\omega'}\right) a . \tag{9.30}$$

Inserting this back into Equation (9.27) yields finally

$$x(t) = a\left[\cos\omega' t \cos\Omega t + \left(\frac{\Omega}{\omega'}\right)\sin\omega' t \sin\Omega t\right] , \tag{9.31}$$

$$y(t) = a\left[-\cos\omega' t \sin\Omega t + \left(\frac{\Omega}{\omega'}\right)\sin\omega' t \cos\Omega t\right] . \tag{9.32}$$

Now $\omega_0 = (g/l)^{1/2} \sim 1\ \mathrm{rads}^{-1}$, for a pendulum of length 10 m. Comparing this with the rotation rate of the Earth ($\omega = 7.27 \times 10^{-5}\ \mathrm{rads}^{-1}$), we see that $\Omega = 2\omega \sin\lambda \ll \omega' = \left(\Omega^2 + \omega_0^2\right)^{1/2}$, whatever the latitude, λ. Consequently we have approximately

$$x(t) \approx a \cos\omega_0 t \cos\Omega t , \tag{9.33}$$
$$y(t) \approx -a \cos\omega_0 t \sin\Omega t . \tag{9.34}$$

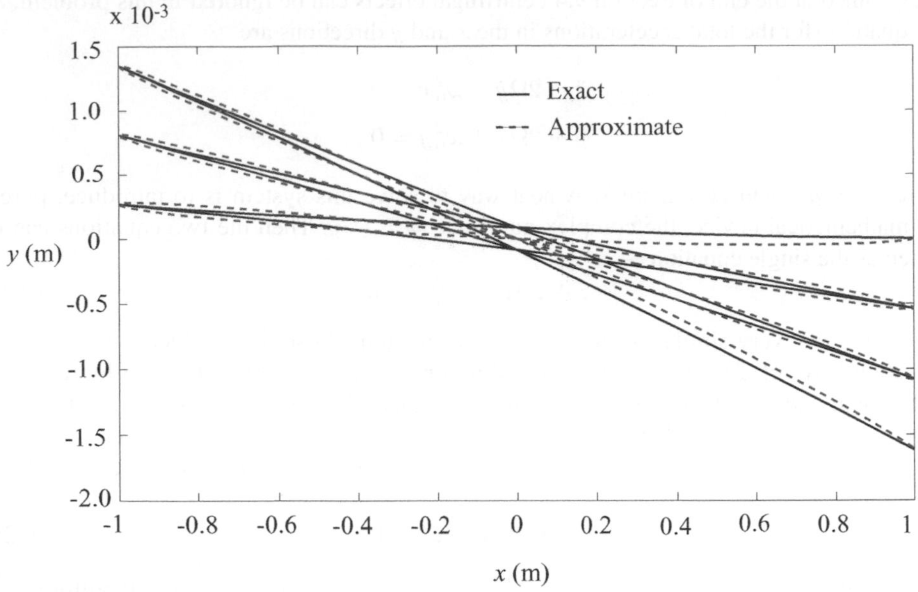

Figure 9.9 Foucault pendulum trajectory in the x-y plane. Both the exact and approximate small amplitude trajectories are shown (Equations (9.31), (9.32) and (9.33), (9.34) respectively). The pendulum was assumed to be 22 m long and located in a laboratory at 52° North.

Both the exact and the approximate solutions are plotted in Figure 9.9 for a pendulum that is 22 m long situated in a laboratory at 52° North The effect of the Coriolis acceleration is to gradually rotate the plane of swing of the pendulum as it executes the relatively rapid oscillations with approximate period $2\pi/\omega_0$. In the northern hemisphere the plane of swing is therefore observed to rotate clockwise when viewed from above. At either pole ($\lambda = \pm 90°$) the plane of swing takes 24 h to complete one revolution. At a general latitude the time for one rotation is given by

$$T = \frac{2\pi}{\Omega} = \frac{2\pi}{\omega \sin\lambda} = \frac{24}{\sin\lambda}\,\text{h}\,. \tag{9.35}$$

The Foucault pendulum in the Science Museum in London (latitude $\lambda \approx 51°$ North), near my office at Imperial College, is 22.45 m long (see Figure 9.10). Its period is $2\pi(l/g)^{1/2} = 9.5$ s and it takes about $24/\sin 51° \approx 30.5$ h to turn through a full circle.

One final remark on the comparison between the exact solution of Equations (9.31) and (9.32) and the approximate solution of Equations (9.33) and (9.34). Whilst the approximate solution correctly gives the initial displacement ($x = a, y = 0$), it incorrectly gives the initial velocity as ($\dot{x}(0) = 0, \dot{y}(0) = -a\Omega$). The small sidewise velocity incorrectly predicts that initially the pendulum bob deviates to the *left* as it approaches the origin for the first time. The accurate solution, however, satisfies the initial conditions and always gives the correct rightwards deflection for Coriolis deviations in the northern hemisphere. Figure 9.11 shows a detailed comparison between the two solutions for the first oscillation period of Figure 9.9. Note from Figs. 9.9 and 9.11 that both solutions agree at the cusp points (reached after

Figure 9.10 The Foucault pendulum at the Science Museum in London. The pendulum's length is 22.45 m.

each pendulum half period), when the extra terms in Equations (9.31) and (9.32), that are not present in Equations (9.33) and (9.34), vanish.

9.7 Free Rotation of a Rigid Body – Tennis Rackets and Matchboxes

Our final example pulls together concepts from Chapter 3 on oscillations, Chapter 8 on rigid bodies, and the current chapter on rotating frames. We consider the most general rigid body of the previous chapter (i.e. it has principal moments $I_1 \neq I_2 \neq I_3$), but unlike there the rotation axis is no longer considered to be fixed. We imagine that referred to the principal axes the body's angular velocity is given by $\boldsymbol{\omega} = (\omega_1, \omega_2, \omega_3)$. Now remember that the principal axes are fixed to the body and rotate with it, so that when referring the analysis to the principal axes of a freely rotating body we are necessarily dealing with a rotating frame. We can apply Equation (9.12) to work out how the rate of change of the (total) angular momentum in the lab frame is related to the rate of change of the angular momentum in a frame that (instantaneously) rotates with the body:

$$\left(\frac{d}{dt}\right)_I \mathbf{L} = \left(\frac{d}{dt}\right)_{\Re} \mathbf{L} + \boldsymbol{\omega} \times \mathbf{L} \,. \tag{9.36}$$

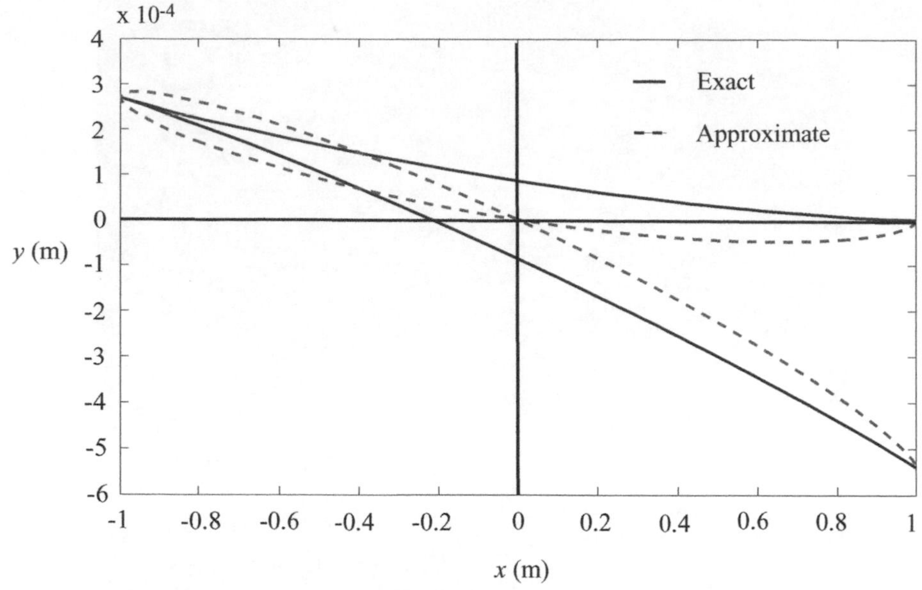

Figure 9.11 First period of Foucault pendulum trajectory of Figure 9.9.

The simplest situation to consider is when the torque, equal to the rate of change of angular momentum in the lab frame, is zero. Using $\mathbf{L} = \mathbf{I}\boldsymbol{\omega}$ we then have in principal coordinates

$$I_1 \frac{d\omega_1}{dt} = \omega_2 \omega_3 (I_2 - I_3) , \tag{9.37}$$

$$I_2 \frac{d\omega_2}{dt} = \omega_3 \omega_1 (I_3 - I_1) , \tag{9.38}$$

$$I_3 \frac{d\omega_3}{dt} = \omega_1 \omega_2 (I_1 - I_2) , \tag{9.39}$$

which are known as *Euler's equations*. Notice that applying Equation (9.12) to $\boldsymbol{\omega}$ itself leaves $\boldsymbol{\omega}$ unchanged and, as a consequence, there is no need to distinguish between the description of $\boldsymbol{\omega}$ between the lab frame and the rotating frame. Now imagine that the body is rotating mainly about principal axis 3, so that initially $\boldsymbol{\omega} = \omega(\delta_1, \delta_2, 1)$, where $\delta_{1,2} \ll 1$. Working to first order in small quantities, we see from Equation (9.39) that $\omega_3 = \omega$ is approximately constant, and that the remaining two equations then become

$$I_1 \frac{d\delta_1}{dt} = (I_2 - I_3)\, \omega \delta_2 , \tag{9.40}$$

$$I_2 \frac{d\delta_2}{dt} = (I_3 - I_1)\, \omega \delta_1 . \tag{9.41}$$

By differentiating we obtain

$$\frac{d^2\delta_1}{dt^2} - q^2\delta_1 = 0 , \tag{9.42}$$

$$\frac{d^2\delta_2}{dt^2} - q^2\delta_2 = 0 , \tag{9.43}$$

where

$$q^2 = \frac{(I_3 - I_1)(I_2 - I_3)}{I_1 I_2}\omega^2 . \tag{9.44}$$

Everything now depends on the sign of q^2. If either $I_3 < I_{1,2}$, or $I_3 > I_{1,2}$, then $q^2 < 0$ and we know from oscillator theory that an imaginary value of q means that the solutions for $\delta_{1,2}$ will both be oscillatory; they will neither grow or decay with time. The system will continue to rotate about the 3-axis in a stable fashion. The only other case to consider is when I_3 lies between the other two principal moments, say $I_1 < I_3 < I_2$. In that case $q^2 > 0$, and the solutions will either grow or decline exponentially with time. Since the general solution is a linear combination of solutions that grow and decline exponentially, we can assert that as time progresses the small perturbations $\delta_{1,2}$ will grow, at least until the approximation $\delta_{1,2} \ll 1$ is no longer valid. The system is *unstable* to rotation about this intermediate rotation axis. The instability is most pronounced when I_3 is significantly distinct from $I_{1,2}$. This is true of a tennis racket, which is why our conclusion is sometimes known as the 'tennis racket theorem'. If you do not have a tennis racket to hand, test the theorem by throwing a matchbox into the air. Each time you throw it up, set it spinning about one of its principal axes (we showed in Section 8.9 that the principal axes of a cuboid are aligned with its sides). You should find that only for the intermediate axis does the box tumble wildly. This is the principle by which satellites are 'spin-stabilised'. In fact, of the two stable axes, the one with the largest principal moment is the most stable, and by setting a satellite to rotate about this axis it maintains its spin orientation.[7]

9.8 Final Thoughts

The kind of rotating frame we have been considering has been restricted to one with constant angular velocity, both in magnitude (i.e. no angular acceleration) and direction (i.e. no precession, nutation, etc.). Although the usefulness of generalising further is doubtful, it does raise an interesting conceptual issue. What is so special about the 'constant velocity' of inertial frames with which we started the discussion of mechanics in this book? What is the most general transformation between reference frames? What indeed *is* a reference frame? A mathematically rigorous discussion would involve us in the theory of differentiable manifolds, but if we accept without scrutiny the notion of an arbitrary frame of reference, then the most pertinent question is: 'Can the laws of mechanics be reformulated in such a way that they *take the same form* in an arbitrary frame of reference?' The answer is 'yes' at both levels at which the question might be treated. It is possible to write down Newtonian mechanics posed in arbitrary *spatial* reference frames, whilst keeping time as a

[7] I once demonstrated this effect in a lecture by tossing a sealed box file. A colleague who saw the demonstration (but not the preceding lecture!) was surprised to find out that I hadn't 'weighted' the box in some way to produce the irregular tumbling when I tossed the box about its intermediate axis.

parameter ticking uniformly, externally and independently of the chosen reference frame. In short, Newtonian mechanics can be reformulated as a consistent scheme *without* recourse to inertial reference frames. It is mathematically difficult, but it can be done. Of course even this description must lead to inconsistencies as speeds approach that of light. The question can therefore be posed again, but now recognising that the 'space' (=manifold) must include time as an *internal* feature of its description. One can then explore generalised transformations (dilations, rotations, velocities, etc.) within this arena, and thereby achieve our most complete description of macroscopic physics. It turns out that gravity can be incorporated very naturally in this generalised description, the result being known as General Relativity. These programmes, which of course take us way outside the domain of this elementary text, and represent one of science's most outstanding achievements, would seem a befitting thought on which to end this book.

9.9 Summary

Section 9.3. The relationship between acceleration, velocity and position as measured in a frame rotating with constant angular velocity $\boldsymbol{\omega}$, as compared with the acceleration in an inertial frame (Figure 9.4) is

$$\mathbf{a}_I = \mathbf{a}_{\Re} + 2\boldsymbol{\omega} \times \mathbf{v}_{\Re} + \boldsymbol{\omega} \times (\boldsymbol{\omega} \times \mathbf{r}) \tag{9.15}$$

where the subscripts I and $\Re$ indicate quantities measured in the inertial frame and rotating frame respectively. The second and third terms on the right are called the **Coriolis acceleration** and the **centrifugal acceleration** respectively.

Section 9.5. When working in a rotating frame attached to the Earth at latitude λ, the Earth's angular velocity is given by

$$\boldsymbol{\omega} = \omega \left(\cos\lambda \mathbf{j} + \sin\lambda \mathbf{k}\right) \ , \tag{9.19}$$

where $\mathbf{j}$ points north, and $\mathbf{k}$ is vertically up. If $\mathbf{v} = v_x\mathbf{i} + v_y\mathbf{j}$ is the velocity in this frame of an object in the plane tangent to the Earth's surface, the Coriolis acceleration is given by

$$2\mathbf{v} \times \boldsymbol{\omega} = 2\omega \left[\sin\lambda \left(v_y\mathbf{i} - v_x\mathbf{j}\right) + v_x \cos\lambda \mathbf{k}\right] \ .$$

Section 9.6. The period of planar rotation of a Foucault pendulum suspended above the Earth's surface at latitude λ is given by

$$T = \frac{2\pi}{\Omega} = \frac{2\pi}{\omega \sin\lambda} = \frac{24}{\sin\lambda}\,\mathrm{h}\,. \tag{9.35}$$

Section 9.7. A freely rotating body with principal moments $I_1 < I_2 < I_3$ is most stable when rotating about the axis with principal moment I_3, and unstable when rotating about the axis with principal moment I_2.

9.10 Problems

9.1 A 500 tonne train runs south at 100 km h^{-1} at a latitude 60° N.

(a) Calculate the (sideways) horizontal force on the tracks.

(b) What is the direction of the force?

9.2 A gun fires shells with a muzzle speed of 500 ms^{-1} and at an angle of 45° to the horizontal. It fires due north from a position at a latitude of 45° N. Estimate the sideways acceleration on the shells when:

(a) at the top of the trajectory;

(b) at the exit from the muzzle;

(c) at the end of the trajectory.

Use the sideways acceleration at the top of the trajectory to estimate by how much the shell will miss a target (25 km away) if the gun, corrected for use at a latitude of 45° N, is used at 45° S instead.

9.3 A person walks with a speed of 1 m s^{-1} from the centre to the edge of a horizontal circular platform which is rotating about a vertical axis through its centre at an angular velocity of 1 rad s^{-1}. Calculate the magnitude of the resultant horizontal force experienced by the person (expressed as a fraction of their weight) due to the rotation when they are 4 m from the centre.

9.4 A bead of mass m slides *without friction* on a rigid rod which rotates in the horizontal plane at a constant angular speed ω about a vertical axis at one end (see Figure 9.12). Derive the equation of motion for r, the distance of the bead from the axis, using a

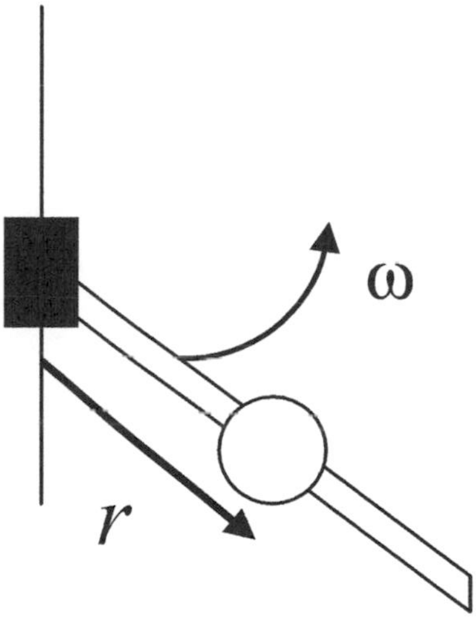

Figure 9.12 Figure for Problem 9.4.

frame of reference in which the rod is stationary, and show (by substitution) that it is satisfied by the expression

$$r = A\exp(\omega t) + B\exp(-\omega t) .$$

By considering the Coriolis force, obtain an expression for the force exerted by the rod on the bead as a function of time (neglect the effects of gravity).

9.5 For this question you may assume that the Earth is a *sphere* of radius $= 6.38 \times 10^6$ m.

(a) A parachutist on a particular jump at latitude $\lambda = 50°$N reaches terminal velocity (6.5 m s^{-1}) when his height above the Earth's surface is 1 km. Ignoring the variation of gravity with height, and the effects of wind, calculate the lateral deviation when he reaches the ground due to (i) the $-2\boldsymbol{\omega} \times \mathbf{v}_I$ term and (ii) the $-\boldsymbol{\omega} \times (\boldsymbol{\omega} \times \mathbf{r})$ term in Equation (9.15). In each case give both the magnitude *and* the direction (e.g. due West).

(b) Will he land inside a target of radius 150 m if initially he aims for the centre of the target area?

(c) On a day when the air is less dense, the terminal velocity attained is greater than that given above. Will this increase or decrease the lateral ground deviation? Explain your answer carefully.

9.6 (A bit tricky!) A mass moving in a straight line on the Earth's surface must be subject to a real force to balance the Coriolis force. The UK stretches from latitude 50° N to 58.5° N. If a uniform westerly wind blows at 20 km h^{-1} across the entire country, estimate the pressure difference between north and south required to maintain the weather pattern. (Density of air $= 1.3$ kg m^{-3}, Earth's radius, $R_E = 6400$ km.)

Appendix 1: Vectors, Matrices and Eigenvalues

In the last three chapters of this book, we discuss effects which are exclusively the domain of mechanics in more than one dimension wherein the equations are necessarily expressed in vector notation. Some quantities, such as work (Section 7.2), torque and angular momentum (Section 7.3) are conveniently expressed in terms of vector products. Matrices and eigenvalues are used in discussing the angular momentum of rigid bodies (Sections 8.7 and 8.8). These topics are briefly reviewed here.

A.1 The Scalar (Dot) Product

Take two vectors **a** and **b** as in Figure A.1. The scalar product between **a** and **b**, denoted

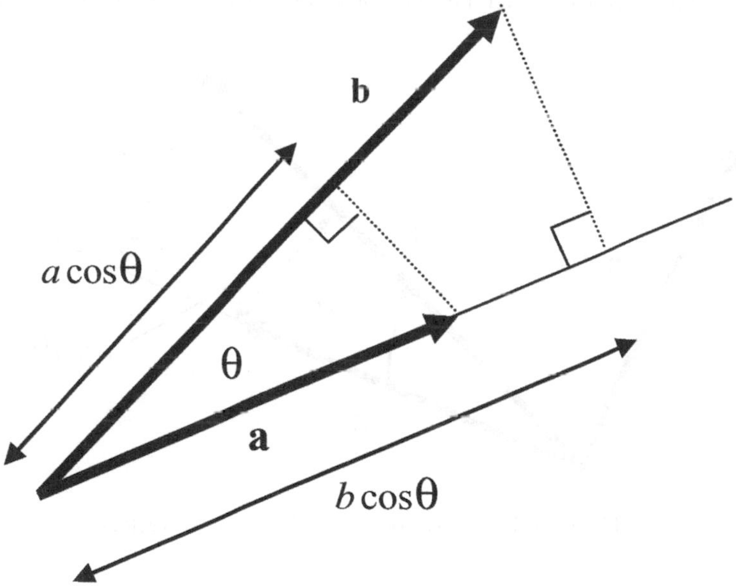

Figure A.1 The scalar product between two vectors.

Classical Mechanics – Second Edition From Newton to Einstein: A Modern Introduction Martin W. McCall

$\mathbf{a} \cdot \mathbf{b}$, is defined as

$$\mathbf{a} \cdot \mathbf{b} = |\mathbf{a}||\mathbf{b}| \cos\theta \,, \tag{A.1}$$

where θ is the angle between $\mathbf{a}$ and $\mathbf{b}$. From the diagram this is seen to be the projection of $\mathbf{a}$ along $\mathbf{b}$ times the magnitude of $\mathbf{a}$ or, since Equation (A.1) is symmetric in $\mathbf{a}$ and $\mathbf{b}$, the projection of $\mathbf{b}$ along $\mathbf{a}$ times the magnitude of $\mathbf{a}$. In Cartesian coordinates $\mathbf{a} \cdot \mathbf{b}$ becomes

$$\mathbf{a} \cdot \mathbf{b} = a_x b_x + a_y b_y + a_z b_z \,. \tag{A.2}$$

When we say the angle between $\mathbf{a}$ and $\mathbf{b}$, there are actually two possibilities: either the angle θ shown (the 'minor' angle), or the angle $2\pi - \theta$ (the 'major' angle). For the scalar product it does not matter which we choose since $\cos(2\pi - \theta) = \cos\theta$. The scalar product is sometimes called the dot product.

A.2 The Vector (Cross) Product

Another useful number that can be formed from the vectors $\mathbf{a}$ and $\mathbf{b}$ is $|\mathbf{a}||\mathbf{b}| \sin\theta$. Now it does matter whether we take the minor angle or the major angle as $\sin(2\pi - \theta) = -\sin\theta$. One way to be consistent with the signs is to agree to always choose the minor angle and then associate a direction with the product $|\mathbf{a}||\mathbf{b}| \sin\theta$. The direction can be assigned by rotating the vector $\mathbf{a}$ towards the vector $\mathbf{b}$ as in Figure A.2 and then use the right-hand rule (see Figure 8.4). The resultant vector is called the cross product between $\mathbf{a}$ and $\mathbf{b}$ and is denoted $\mathbf{a} \times \mathbf{b}$.

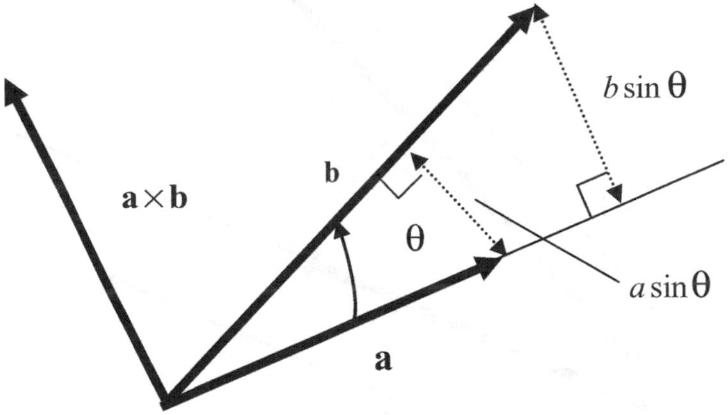

Figure A.2 The vector product between two vectors.

Geometrically, the quantity $|\mathbf{a}||\mathbf{b}| \sin\theta$ is the perpendicular distance of $\mathbf{a}$ from $\mathbf{b}$ times the magnitude of $\mathbf{b}$, or the perpendicular distance of $\mathbf{b}$ from $\mathbf{a}$ times the magnitude of $\mathbf{a}$. In Cartesian coordinates the cross product is given by

$$\mathbf{a} \times \mathbf{b} = (a_y b_z - a_z b_y)\,\mathbf{i} - (a_x b_z - a_z b_x)\,\mathbf{j} + (a_x b_y - a_y b_x)\,\mathbf{k} \,. \tag{A.3}$$

Note the cross product is antisymmetric: $\mathbf{a} \times \mathbf{b} = -\mathbf{b} \times \mathbf{a}$.

A.3 The Vector Triple Product

Since the cross product between two vectors $\mathbf{b}$ and $\mathbf{c}$ is a vector, we can take another cross product with another vector (say $\mathbf{a}$) to produce yet another vector:

$$\mathbf{a} \times (\mathbf{b} \times \mathbf{c}) \ .$$

(Note: the brackets are *essential*!) From the way the cross product is constructed, we see from

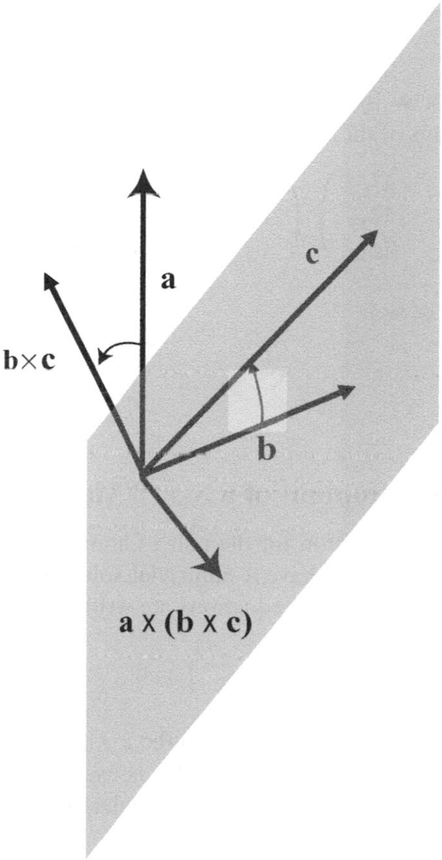

Figure A.3 The vector triple product between three vectors.

Figure A.3 that the vector $\mathbf{a} \times (\mathbf{b} \times \mathbf{c})$ must lie in the plane of $\mathbf{b}$ and $\mathbf{c}$. This means that it must be possible to write $\mathbf{a} \times (\mathbf{b} \times \mathbf{c})$ in the form $\beta\mathbf{b} + \gamma\mathbf{c}$, where β and γ are numbers. In fact

$$\mathbf{a} \times (\mathbf{b} \times \mathbf{c}) = (\mathbf{a} \cdot \mathbf{c})\, \mathbf{b} - (\mathbf{a} \cdot \mathbf{b})\, \mathbf{c} \ , \tag{A.4}$$

as you can verify by writing out the components in Cartesian coordinates.

A.4 Multiplying a Vector by a Matrix

Suppose we have a square 3×3 matrix given by

$$\mathbf{M} = \begin{pmatrix} M_{11} & M_{12} & M_{13} \\ M_{21} & M_{22} & M_{23} \\ M_{31} & M_{32} & M_{33} \end{pmatrix}, \tag{A.5}$$

and a vector $\mathbf{v}$, whose Cartesian components are represented by the column

$$\mathbf{v} = \begin{pmatrix} v_1 \\ v_2 \\ v_3 \end{pmatrix}. \tag{A.6}$$

We can generate a new vector $\mathbf{w}$ by multiplying the vector into the matrix. Denoting this product as $\mathbf{Mv}$, the components of the new vector are given by

$$\mathbf{w} = \mathbf{Mv} = \begin{pmatrix} M_{11} & M_{12} & M_{13} \\ M_{21} & M_{22} & M_{23} \\ M_{31} & M_{32} & M_{33} \end{pmatrix} \begin{pmatrix} v_1 \\ v_2 \\ v_3 \end{pmatrix} = \begin{pmatrix} M_{11}v_1 + M_{12}v_2 + M_{13}v_3 \\ M_{21}v_1 + M_{22}v_2 + M_{23}v_3 \\ M_{31}v_1 + M_{32}v_2 + M_{33}v_3 \end{pmatrix}. \tag{A.7}$$

Generally, the ith component of $\mathbf{w}$ is

$$w_i = \sum_{j=1}^{3} M_{ij} v_j \,. \tag{A.8}$$

A.5 Calculating the Determinant of a 3 × 3 Matrix

In Section 3.13 we saw that the condition for the pair of homogeneous simultaneous linear equations $ax + by = 0; cx + dy = 0$ to have a nontrivial solution was that the coefficients must satisfy $ad - bc = 0$. When the pair of equations are written in the matrix form

$$\begin{pmatrix} a & b \\ c & d \end{pmatrix} \begin{pmatrix} x \\ y \end{pmatrix} = 0\,, \tag{A.9}$$

we see that the condition amounts to an operation on the 2×2 matrix to produce a number. This operation is called finding the determinant, and it is useful to extend the definition to matrices of higher order. We will only go up one to calculate the determinant of the 3×3 matrix

$$\begin{pmatrix} M_{11} & M_{12} & M_{13} \\ M_{21} & M_{22} & M_{23} \\ M_{31} & M_{32} & M_{33} \end{pmatrix}. \tag{A.10}$$

First set up a chequer board of signs:

$$\begin{pmatrix} + & - & + \\ - & + & - \\ + & - & + \end{pmatrix}. \tag{A.11}$$

Now, from the original matrix, in your imagination cross off the row and column that both contain the element M_{11}, and calculate the determinant of the resulting 2×2 matrix as

$(M_{22}M_{33} - M_{23}M_{32})$. Multiply this number by the element M_{11} itself, but with a sign given by the sign occupying the location of M_{11} in the chequer board (A.11), to form

$$D_1 = +M_{11}\,(M_{22}M_{33} - M_{23}M_{32}) \ . \tag{A.12}$$

Now move onto the element M_{12} and repeat the procedure to arrive at

$$D_2 = -M_{12}(M_{21}M_{33} - M_{23}M_{31}) \, . \tag{A.13}$$

Finally, for M_{13}

$$D_3 = +M_{13}(M_{21}M_{32} - M_{22}M_{31}) \, . \tag{A.14}$$

The determinant, denoted det $\mathbf{M}$, is given by the sum of all three terms, that is

$$\begin{aligned} \det \mathbf{M} &= D_1 + D_2 + D_3 \\ &= +M_{11}M_{22}M_{33} - M_{11}M_{23}M_{32} \\ &\quad -M_{12}M_{21}M_{33} + M_{12}M_{23}M_{31} \\ &\quad +M_{13}M_{21}M_{32} - M_{13}M_{22}M_{31} \, . \end{aligned} \tag{A.15}$$

Notice that although we worked along the top row in forming the determinant we could, in fact, have worked along *any* row or column, and achieved the same result. The extension to the general definition of the determinant of an $N \times N$ matrix is straightforward, but will not be given here. Have a guess at what it is, before checking on Wikipedia.

A.6 Eigenvectors and Eigenvalues

Changing the numerical subscripts to Cartesian labels, the matrix of the previous section becomes

$$\mathbf{M} = \begin{pmatrix} M_{xx} & M_{xy} & M_{xz} \\ M_{yx} & M_{yy} & M_{yz} \\ M_{zx} & M_{zy} & M_{zz} \end{pmatrix} . \tag{A.16}$$

We can interpret such an array as an ‘operator’ acting on Cartesian vectors $\mathbf{v} = v_x\mathbf{i} + v_y\mathbf{j} + v_z\mathbf{k}$ to produce a new vector according to

$$\mathbf{Mv} = \begin{pmatrix} M_{xx}v_x + M_{xy}v_y + M_{xz}v_z \\ M_{xy}v_x + M_{yy}v_y + M_{yz}v_z \\ M_{xz}v_x + M_{yz}v_y + M_{zz}v_z \end{pmatrix} , \tag{A.17}$$

where the matrix product was defined in Section A.4. Now it may happen that the effect of the operator on a particular vector $\mathbf{e}$ is to simply multiply it by a scalar, λ, so that

$$\mathbf{Me} = \lambda\mathbf{e} \, . \tag{A.18}$$

When this happens we call $\mathbf{e}$ an *eigenvector* of $\mathbf{M}$, and λ its corresponding *eigenvalue*. How do we search for the eigenvectors and eigenvalues of a given matrix systematically? First recast Equation (A.18) as

$$(\mathbf{M} - \lambda\mathbf{Id})\,\mathbf{e} = \mathbf{0} \, , \tag{A.19}$$

where $\mathbf{Id}$ represents the *identity operator*, that is the operator that leaves any vector unchanged: $\mathbf{Id}(\mathbf{v}) = \mathbf{v}$. The zero on the right-hand side of Equation (A.19) is in bold because it is actually the null vector $\mathbf{0} = 0\mathbf{i} + 0\mathbf{j} + 0\mathbf{k}$. Equation (A.19) is a set of homogeneous simultaneous equations. By analogy with the 2×2 case (cf. Sections 3.13 and A.5) the condition for the existence of a nontrivial solution is that the determinant of $(\mathbf{M} - \lambda\mathbf{Id})$ should vanish, that is

$$\det(\mathbf{M} - \lambda\mathbf{Id}) = 0\,. \tag{A.20}$$

For the 3×3 case this will be a cubic equation in λ. Let's illustrate with a specific example. Set

$$\mathbf{M} = \begin{pmatrix} 1 & 0 & 0 \\ 0 & 1 & 4 \\ 0 & 1 & -2 \end{pmatrix} \tag{A.21}$$

The determinant condition is

$$\det\left[\begin{pmatrix} 1 & 0 & 0 \\ 0 & 1 & 4 \\ 0 & 1 & -2 \end{pmatrix} - \begin{pmatrix} \lambda & 0 & 0 \\ 0 & \lambda & 0 \\ 0 & 0 & \lambda \end{pmatrix}\right] = 0\,, \tag{A.22}$$

or

$$(1-\lambda)\left(\lambda^2 + \lambda - 6\right) = 0\,, \tag{A.23}$$

or

$$(1-\lambda)(\lambda - 2)(\lambda + 3) = 0\,. \tag{A.24}$$

(I chose the numbers so it would factorize easily.) The three eigenvalues are therefore $\lambda = 1, \lambda = 2$ and $\lambda = -3$. Associated with each of these is a corresponding eigenvector. Let us find the eigenvector $\mathbf{e} = e_x\mathbf{i} + e_y\mathbf{j} + e_z\mathbf{k}$, associated with $\lambda = 2$. We see from Equation (A.18) that $\lambda = 2$ requires

$$\begin{pmatrix} 1 & 0 & 0 \\ 0 & 1 & 4 \\ 0 & 1 & -2 \end{pmatrix}\begin{pmatrix} e_x \\ e_y \\ e_z \end{pmatrix} = 2\begin{pmatrix} e_x \\ e_y \\ e_z \end{pmatrix}\,. \tag{A.25}$$

The first line, $e_x = 2e_x$, is only true if $e_x = 0$. The second and third lines both yield $e_y = 4e_z$, so a possible eigenvector is $\mathbf{e} = 4\mathbf{j} + \mathbf{k}$. Any scalar multiple of this vector is also an eigenvector (with the same eigenvalue), so an eigenvector really defines a particular *direction* in which the operator acts as in Equation (A.18). Using a similar procedure you should be able to show that $\mathbf{j} - \mathbf{k}$ is an eigenvector associated with $\lambda = -3$. For $\lambda = 1$ we require

$$\begin{pmatrix} 1 & 0 & 0 \\ 0 & 1 & 4 \\ 0 & 1 & -2 \end{pmatrix}\begin{pmatrix} e_x \\ e_y \\ e_z \end{pmatrix} = \begin{pmatrix} e_x \\ e_y \\ e_z \end{pmatrix}\,. \tag{A.26}$$

The second and third lines are $e_y + 4e_z = e_y$ and $e_y - 2e_z = e_z$, which is only true if $e_y = e_z = 0$. The first line implies that e_x is arbitrary, so an eigenvector is $\mathbf{e} = \mathbf{i}$. Collecting together, we have the three eigenvectors with their corresponding eigenvalues as

$$\lambda_1 = 1 \leftrightarrow \mathbf{e}_1 = \begin{pmatrix} 1 \\ 0 \\ 0 \end{pmatrix}\;;\; \lambda_2 = 2 \leftrightarrow \mathbf{e}_2 = \begin{pmatrix} 0 \\ 4 \\ 1 \end{pmatrix}\;;\; \lambda_3 = -3 \leftrightarrow \mathbf{e}_3 = \begin{pmatrix} 0 \\ 1 \\ -1 \end{pmatrix}\,. \tag{A.27}$$

Sometimes it is convenient to rescale the eigenvectors so that they have unit length. For the current case we could then set

$$\mathbf{e}_1 = \begin{pmatrix} 1 \\ 0 \\ 0 \end{pmatrix}, \ \mathbf{e}_2 = \frac{1}{\sqrt{17}} \begin{pmatrix} 0 \\ 4 \\ 1 \end{pmatrix}, \ \mathbf{e}_3 = \frac{1}{\sqrt{2}} \begin{pmatrix} 0 \\ 1 \\ -1 \end{pmatrix}, \tag{A.28}$$

to achieve a set of three unit eigenvectors. This detailed worked example should enable you to calculate the eigenvalues and eigenvectors of simple matrices. The same technique can be applied to larger matrices, though finding the roots of higher-order polynomials, and the algebra associated with finding the eigenvectors, rapidly becomes unwieldy. Note that it is possible for an eigenvalue to be zero and the corresponding eigenvector to be nonnull. If some eigenvalues are the same, then we must be careful not to overlook an eigenvector. If, say, $\lambda_1 = \lambda_2$ then there are still two distinct directions associated with each eigenvalue. This situation is referred to as a *degeneracy*. Note also that even when the elements of the original matrix are all real, it is possible for the eigenvalues to be complex numbers, though the interpretation of this situation need not concern us.

A.7 Diagonalising Symmetric Matrices

Suppose it happens that the elements of a matrix satisfy $M_{ij} = M_{ji}\,,\ i,j = 1,2,3$, so that for the matrix of the previous section we have

$$\mathbf{M} = \begin{pmatrix} M_{xx} & M_{xy} & M_{xz} \\ M_{xy} & M_{yy} & M_{yz} \\ M_{xz} & M_{yz} & M_{zz} \end{pmatrix}. \tag{A.29}$$

Such a matrix is referred to as *symmetric*. Symmetric matrices have the property that for any two vectors $\mathbf{u}$ and $\mathbf{v}$

$$\mathbf{u} \cdot (\mathbf{M}\mathbf{v}) = (\mathbf{M}\mathbf{u}) \cdot \mathbf{v}\,, \tag{A.30}$$

which you can verify by writing out the components. Now let us suppose that $\mathbf{u}$ and $\mathbf{v}$ are both eigenvectors ($\mathbf{e}_1 = \mathbf{u}$, $\mathbf{e}_2 = \mathbf{v}$) associated with distinct eigenvalues $\lambda_1 \neq \lambda_2$. It is then clear from Equation (A.30) that

$$\mathbf{e}_1 \cdot (\mathbf{M}\mathbf{e}_2) = \lambda_2 \mathbf{e}_1 \cdot \mathbf{e}_2 = (\mathbf{M}\mathbf{u}) \cdot \mathbf{v} = \lambda_1 \mathbf{e}_1 \cdot \mathbf{e}_2\,. \tag{A.31}$$

This implies that $\mathbf{e}_1 \cdot \mathbf{e}_2 = 0$, since $\lambda_1 \neq \lambda_2$. For the three eigenvectors of $\mathbf{M}$ we could have chosen any pair and obtained the same result, namely that the pair are orthogonal. We therefore have the important conclusion that for a matrix $\mathbf{M}$ with distinct eigenvalues, the corresponding eigenvectors are orthogonal. Should degeneracy occur (e.g. $\lambda_1 = \lambda_2$), then it turns out that the corresponding eigenvectors can be *chosen* to be orthogonal. This means that associated with any $\mathbf{M}$ is a set of three orthogonal directions. If we choose these three directions to be our coordinate axes ($\mathbf{i} = \mathbf{e}_1, \mathbf{j} = \mathbf{e}_2, \mathbf{k} = \mathbf{e}_3$) then the matrix $\mathbf{M}$ must take the simple form

$$\begin{pmatrix} \lambda_1 & 0 & 0 \\ 0 & \lambda_2 & 0 \\ 0 & 0 & \lambda_3 \end{pmatrix}, \tag{A.32}$$

since then it acts on each unit eigenvector to multiply it by its corresponding eigenvalue. Finding the orthogonal eigendirections associated with $\mathbf{M}$ and rotating the original coordinates to be aligned with these directions, so that $\mathbf{M}$ takes the simple form of Equation (A.32), is called *diagonalising* $\mathbf{M}$.

For ease of visualization we have illustrated diagonalisation in three dimensions. However, the procedure can be applied to square matrices of any dimension provided, of course, that the matrix is symmetric. We will not prove it here, but symmetry also implies that the eigenvectors are all real.

Appendix 2: Answers to Problems

Full worked solutions may be obtained via www.wiley.com/go/mccall.

Chapter 2

2.1 40 km h^{-1}.

2.2 (a) $\sqrt{12}$ s, (b) $20\sqrt{3}$ m s^{-1}, (c) $300\sqrt{3}$ m s^{-1}, (d) 15 m.

2.3 At $t = 0$ s and $t = \pi$ s : (i) 2 m s^{-1} in the $+y$ direction, (ii) 4 m s^{-2} in the $-x$ direction. At $t = \pi/2$ s : (i) 2 m s^{-1} in the $-y$ direction, (ii) 4 m s^{-2} in the $+x$ direction. SHM with period π s.

2.4 (a) $2\sqrt{61}$ m s^{-1}, (b) $(2\mathbf{i} - 24\mathbf{j})$ m s^{-2}, (c) $(6\mathbf{i} - 12\mathbf{j})$ m s^{-2}.

2.5 $(10\mathbf{i} + \mathbf{j})$ m s^{-1}.

2.6 25 m s^{-1}, $-20\mathbf{k}$ N. Parabolic.

2.7 (a) 3 J, (b) -0.6 J, (c) 2.4 J, (d) 3 W.

2.8 500 N, 30 kW, 5%.

2.9 $h = m^2u^2/[2g(m + M)^2]$.

2.10 60 N, 32 s, 64 m s^{-1}.

2.11 (i) $a = (m_2 - m_3)g/(m_2 + m_3)$, (ii) $T = 2m_2m_3g/(m_2 + m_3)$. a and T are doubled, balance remains horizontal.

2.12 (i) 6.6 m s^{-1}, (ii) 9.3 m s^{-1}.

2.13 (c) $\pi/4$, $R_{\text{max}} = 40.8$ m, (d) 7.2 m s^{-1}, (f) Batsman will not be caught.

Chapter 3

3.1 4π rad s^{-1}, 2 Hz, 0.5 s.

3.2 400 N m^{-1}.

3.3 (i) $\sqrt{2}$ cm, 0.027 J, (iii) 0.42 m s^{-1}.

3.4 (a) 4, (b) 64, (c) 2.0005 s.

3.6 1.59 kHz.

3.7 ≈ 122 . Approximately 7 years.

3.8 (c) $x(0) = 0$ m, $\dot{x}(0) = -4$ m s^{-1}, (d) $\phi = +\pi/2$, $A = 0.8$ m, (e) 1.26 s, (f) 4, (g) 0.014 s^{-1}, (h) 361, (i) 10^{-6}.

3.9 (b) 3.4 s, (c) 12.5 J, (d) 167, (e) 116 s, (f) 0.32 Hz, (g) $\pi/4$.

3.10 3.09 m s^{-1}. Induced displacement is ≈ 1.3 m, so the bridge will collapse.

3.11 (a) 10^5 N m^{-1}, (b) 0.63 s, (c) 2×10^4 kg s^{-1}, (d) 5 rad s^{-1}, 5 s^{-1}, (e) 1.45 s, (f) 11.25 m s^{-1}, (g) Faster.

3.12 (f) $\omega_1 = 8.3$ rad s^{-1}, $\omega_2 = 3.4$ rad s^{-1}.

Chapter 4

4.1 0.75 kg, $(-1.5\text{ m}, 0.25\text{m})$.

4.2 (a) $(1.25, 0.75)$ m s^{-1}, (b) $(1.0, 3.0)$ m s^{-1}, (c) $\pm(0.75, 2.25)$ N s, (d) 3.75 J.

4.3 3 s.

4.4 5 m.

4.6 (b) $(-1.33, -1.0)$ m s^{-1}, (c) $(1.33, -2.0)$ m s^{-1}, (g) 4.

4.7 (a) 3.3×10^3 m s^{-1}, (b) 6.7×10^3 m s^{-1}, (c) -3.3×10^3 m s^{-1}, (d) 1.5, (e) 8/9, (f) 8, (g) 30.

4.8 $e^{1000} \approx 10^{434}$ kg (!) This far exceeds the mass of the visible universe.

4.10 (a) (i) 207 kg, (ii) 1770 kg, (iii) Negative, i.e. impossible, (b) First stage: 36219 kg, second stage: 1770 kg, (c) 137 kg s^{-1}, (d) 264 s, (e) 9.3 km s^{-1}.

Chapter 5

5.1 $0.66c = 2 \times 10^8$ m s^{-1}.

5.2 $0.3c = 0.9 \times 10^8$ m s^{-1}.

5.3 $u/c \sim 0.4$.

5.4 $\gamma = 1 + u^2/(2c^2) + \cdots O(u^4/c^4)$.

5.5 $0.4c = 1.2 \times 10^8$ m s^{-1}, 11 years.

5.7 Miss.

5.8 2 years.

5.9 (a) $-0.98c$, (b) $+0.98c$, (c) $1.8c$.

5.10 $t' = -1.26$ s, $t'' = +1.26$ s. None.

5.12 (a) $\gamma = 2$, $u = (\sqrt{3}/2)c = 0.87c$, (b) 5 m, (e) $t_F = 38.5$ ns $= t_B$, $t'_F = 19.3$ ns, $t'_B =$ 77.0 ns, (f) Good.

Chapter 6

6.1 (a) 511 keV c^{-2}, (b) 939 MeV c^{-2}, (c) 1.2 MeV, (d) 532 eV, (e) 532 eV, (f) 145 MeV.

6.2 (a) 8.2×10^{-14} J, (b) 0.16 pJ, (c) 1.6 J.

6.3 (a) 3 nm, (b) 220 nm, , (c) 220 μm.

6.4 $\approx 1.2 \times 10^3$ kg yr^{-1}, (b) 3.2 GJ kg^{-1}.

6.5 (a) $x'_A = -62$ m, $t'_A = 230$ ns, $x'_B = -230$ m, $t'_B = 690$ ns, (b) $x_C = +62$ m, $t_C = 230$ ns, $x_D = -230$ m, $t_D = -690$ ns, (c) $p' = -0.31$ MeV c^{-1}, $E' = 0.59$ MeV, (d) $p = 6.34$ MeV c^{-1}, $E = 6.36$ MeV.

6.6 (a) 67.5 MeV, 67.5 MeV c^{-1}, (b) 249 MeV, 209 MeV c^{-1}.

6.7 1.15 MeV c^{-2}.

6.9 (a) 3.7027 GeV c^{-2}, (b) 3.7074 GeV, 0.18550 GeV c^{-1}, (c) $A = 0.9981$, $B = 0.9987$, (d) 1.25×10^{-3}.

6.10 26.86 MeV.

6.11 4.85.

6.12 £1.34×10^9.

Chapter 7

7.1 480 kg m^2 s^{-1}, 6 m s^{-1}.

7.2 42×10^3 km, three.

7.3 $g_{\text{Moon}} = 0.16 g_{\text{E}} = 1.57$ m s^{-2}, $v_{\text{Moon}}^{\text{escape}} = 0.21 v_{\text{E}}^{\text{Escape}} = 2.38$ km s^{-1}.
$g_{\text{Jupiter}} = 2.63 g_{\text{E}} = 25.8$ m s^{-2}, $v_{\text{Jupiter}}^{\text{escape}} = 5.38 v_{\text{E}}^{\text{Escape}} = 60.2$ km s^{-1}.

7.4 17.94 a.u.

7.5 7×10^{10} Solar masses.

7.6 (a) $\dot{\mathbf{r}} = [-6\sin(2t)\mathbf{i} + 2\cos(2t)\mathbf{j} - 2\sqrt{3}\cos(2t)\mathbf{k}]$ m s^{-1},
$\ddot{\mathbf{r}} = [-12\cos(2t)\mathbf{i} - 4\sin(2t)\mathbf{j} + 4\sqrt{3}\sin(2t)\mathbf{k}]$ m s^{-2} $= -4\mathbf{r}$ m s^{-2},
(b) $[-24\cos(2t)\mathbf{i} - 8\sin(2t)\mathbf{j} + 8\sqrt{3}\sin(2t)\mathbf{k}]$ N $= -8\mathbf{r}$ N.
(c) $[12\sqrt{3}\mathbf{j} + 12\mathbf{k}]$ kg m^2 s^{-1}, (d) **0** N m, (e) Central, (h)(i) 20 J, (h)(ii) 20 J.

7.7 (a) 30 km s^{-1}, (e) $v_1 = 40$ km s^{-1}, $v_2 = 4$ km s^{-1}, (f) 14 km s^{-1}.

Chapter 8

8.1 (a) $ML^2/3$, (b) $2ML^2/5$, (c) $7ML^2/5$, (d) $ML^2/2$, (e) ML^2.

8.2 1.13×10^{11} Hz, 2.66×10^{-3} m, i.e. μ-wave.

8.3 (f) solid cylinder will reach first, 1.15 times faster.

8.4 $Mg/2$.

8.5 (a) $I_1 = [m_1 m_2/(m_1 + m_2)]\, l^2 = \mu l^2$, $I_2 = I_3 = 0$, (b) $I_1 = 2m_1 m_2 l^2 \cos^2\theta/(m_1 + 2m_2)$, $I_2 = 2m_2 l^2 \sin^2\theta$, $I_3 = I_1 + I_2$, asymmetric unless $\tan\theta = (1 + 2m_2/m_2)^{-1/2}$, (c) $I_1 = I_2 = m_2\left[2m_1/(m_1 + 3m_2) + \frac{1}{2}\right] l^2$, $I_3 = m_2 l^2$.

8.6 (a) $I_1 = 10ma^2$, $\mathbf{e}_1 = (1, 0, 0)$; $I_2 = 10ma^2$, $\mathbf{e}_2 = 2^{-1/2}(0, 1, -1)$; $I_3 = 2ma^2$, $\mathbf{e}_3 = 2^{-1/2}(0, 1, 1)$, (b) $233ma^2$ J, (c) $I_2' = 20ma^2$.

8.7 (b) 2.9 kJ, 506 N m.

8.8 (a) 1067 kg m^2, (b) 0.13 rad s^{-1}, (c) 23.8 kN, (e) 335 m.

Chapter 9

9.1 (a) 1749 N, (b) west.

9.2 (a) 3.7×10^{-2} m s^{-2} east, (b) 0, (c) 7.3×10^{-2} m s^{-2} east, 92.5 m.

9.3 0.45.

9.4 $2m\omega^2\,(Ae^{\omega t} - Be^{-\omega t})$.

9.5 (a)(i) 6.08×10^{-4} m east, (ii) 1.66×10^{-2} m south, (b) no, (c) decrease.

9.6 805 N m^{-2}.

Appendix 3: Bibliography

It goes without saying that there are many texts that develop the material here in a more formal way. Below are listed the books that I have enjoyed consulting in the past.

1. **University Physics with Modern Physics**, (12th Edition), by H.D. Young and R.A. Freedman, *Addison-Wesley Publishing Co.*, (2008).

   This has been the recommended first-year text for undergraduates studying physics at Imperial College. Chapters 1-13 cover Newtonian mechanics, and Chapter 37 covers relativity, both at a more elementary level than in this book. A vast number of topical examples and problems are. included . The presentation is pedagogically excellent.

2. **Feynman Lectures On Physics - Volume 1**, by R.P. Feynman, R.B. Leighton and M. Sands, *Addison-Wesley Publishing Co.*, (1963).

   There are few physicists who do not find Feynman's famous lecture series the most inspirational undergraduate text ever written. Feynman's unique charisma shines as brilliantly now as for the undergraduate students he taught in the 1960s. Newtonian mechanics, relativity and much more are covered in the first volume. The second volume contains a chapter on the principle of least action, not discussed here. There are no problems.

3. **Spacetime Physics**, (2nd Edition), by E.F. Taylor and J.A. Wheeler, *W.H.Freeman and Co.*, (1992).

   This is an inspiring text on relativity, with plenty of stimulating discussion, examples and paradoxes. Enthusiasts who really want to get to grips with relativity, beyond passing examinations, should hunt out the first edition of this book, which contains significant concepts omitted in the second.

4. **Classical Mechanics**, (3rd Edition), by H. Goldstein, C.P. Poole Jr and J.L. Safko, *Addison-Wesley Publishing Co.*, (2001).

   An advanced and detailed text, suitable for those wishing to study classical mechanics to a higher level.

5. **Classical Mechanics**, (5th Edition), by T.W.B. Kibble and F.H. Berkshire, *World Scientific Publishing Co.*, (2004).

   Written by colleagues at Imperial College, this book goes beyond the present text, but covers the material very clearly. It includes chapters on dynamical systems and chaos, but not relativity.

Classical Mechanics – Second Edition From Newton to Einstein: A Modern Introduction Martin W. McCall

INDEX

Classical Mechanics – Second Edition From Newton to Einstein: A Modern Introduction Martin W. McCall